EVOLUTIONARY
HUMANISM

EVOLUTIONARY HUMANISM

JULIAN HUXLEY

With an Introduction by
H. James Birx

GREAT MINDS SERIES

PROMETHEUS BOOKS
Buffalo, New York

Published 1992 by Prometheus Books
59 John Glenn Drive, Buffalo, New York 14228,
716-837-2475. FAX: 716-835-6901.

Library of Congress Cataloging-in-Publication Data

Huxley, Julian, 1887–1975.
  [Essays of a humanist]
  Evolutionary humanism / Julian Huxley ; with an introduction by
H. James Birx.
      p.   cm. — (Great minds series)
  Originally published: Essays of a humanist. 1st ed. New York : Harper
and Row, 1964.
  Includes index.
  ISBN 0-87975-778-7
  1. Biology. I. Title. II. Series.
QH311.H83   1992
304.2—dc20   92-33243
                                                                  CIP

Printed in Canada on acid-free paper.

Also Available in Prometheus's
Great Minds Paperback Series

Introduction

SIR JULIAN SORELL HUXLEY, born in London on June 22, 1887, attended Eton and then Balliol College, Oxford. His early interests in experimental biology included genetics, embryology, morphology, physiology, growth and development, and nature conservation.

Huxley's first teaching position was as a professor of biology at Rice University in Texas (1913–1916). In 1925, he was appointed Professor of Zoology at King College, London University. For seven years, he was secretary of the Zoological Society of London (1935–1942) and later became Director-General of UNESCO (1946–1948). Huxley received the Darwin Medal of the Royal Society in 1956, was knighted two years later, and was awarded the Darwin Medal from the German Democratic Republic in 1959.

Julian Huxley was a great scientist, prolific writer, and leading champion of evolutionary thought. As a distinguished naturalist who saw the intimate connection between facts and values, he became a respected spokesman for science and humanism. Furthermore, his own dedication to scientific research did not blind him either to the alarming problems of the human situation or to the need for resolvable action.

Early in his career as a biologist, Huxley committed himself to the fact of evolution and its awesome implications. He saw biology as a link between mind and matter, and held evolution to be the link between science and religion. Huxley recognized the crucial differences between organic evolution in general with its explanatory mechanisms, and human evolution in particular with its sociocultural development. For him, a critical difference separates the ongoing evolution of humankind from material nature (especially in the realm of ethics).

Huxley's early research projects and resultant publications were a major contribution to biology. On two extensive field trips to central East Africa in 1929 and 1960, Huxley studied, with characteristic enthusiasm and energy, the teeming wildlife of the continent (especially large mammals). He became acutely aware of the need to preserve those vanishing species threatened with extinction, and even fought for the worldwide conservation of wildlife and natural habitats. Huxley's book *Africa View* appeared in 1931. He also delighted in bird watching. An avid ornithologist, Huxley knew that the adaptive radiation (the creative fanning out of life forms) in early birds clearly illustrates those results of evolutionary forces, e.g, the speciation of Galapagos finches and Hawaiian honeycreepers. These research projects made Huxley more and more aware of the problems involving ecology and changing environments.

In the nineteenth century, Charles Darwin had fathered the scientific theory of organic evolution grounded in facts and logic.[1] With his two major books, *On the Origin of Species* (1859)[2] and *The Descent of Man* (1871), Darwin's ideas also paved the way for research in ecology, ethnology, and psychology. In writing about Darwin, Julian Huxley pointed out the crucial influences that had helped the young naturalist to develop his scientific theory of organic evolution: the geologist Sir Charles Lyell (1797–1875), the voyage of the HMS *Beagle,* and the economist Thomas R. Malthus (1766–1834). For Huxley, Darwin stands out as the most important evolutionist in the history of biology, for he had taken the idea of evolution in natural philosophy and presented it

---

1. See H. James Birx, *Interpreting Evolution: Darwin & Teilhard de Chardin* (Buffalo, N.Y.: Prometheus Books, 1991), pp. 112–65.

2. Charles Darwin, *The Origin of Species* (Buffalo, N.Y.: Prometheus Books, 1991).

as a scientific theory founded upon the primary explanatory mechanism of natural selection or the survival of the fittest.

Reminiscent of his paternal grandfather Thomas H. Huxley (1825–1895), who eagerly defended the Darwinian theory of organic evolution in the nineteenth century, Julian Huxley advocated the synthetic theory of biological evolution in the middle of this century. Referred to as Neodarwinism, the present synthetic theory of organic evolution is explained in terms of genetic variation and population dynamics as well as Darwinian natural selection.

Like paleontologist George Gaylor Simpson and geneticist Theodosius Dobzhansky (among others), Huxley wrote popular books that helped the general reader to understand and appreciate the fact of evolution and its wide-ranging consequences for science as well as philosophy and theology.

Critical of Arnold Toynbee's limiting interpretation of human history, which saw cultural evolution merely in terms of a few thousand years, Huxley argued for an evolutionary view of the emergence of our species within a cosmic perspective. Also concerned with the future of humankind, Huxley singled out the alarming population growth rate of our species as the most serious world problem now facing all life on this planet.

Unlike many other biologists, Julian Huxley was not reluctant to consider seriously the far-reaching implications that evolution holds for comprehending the place of our species within dynamic nature. As an optimist and humanist, he gave preference to science and reason over the traditional interpretations of humankind and the world grounded merely in myths and beliefs. Furthermore, Huxley boldly called for a new religion of evolution without any appeal to revelation or supernaturalism. He replaced the traditional god-centered religions with an evolution-centered philosophy of humanism. Unlike Huxley, however, the modern secular humanist sees no need to speak at all of divinity within the world.

As an atheist, Huxley called for a new assessment of established religion and pondered the practical applications of the empirical facts supporting organic evolution for the solution of biological and social problems (particularly in the areas of eugenics and ethics). For him, our own species is responsible for the continuing survival of life on Earth and the fulfillment of mental activity in the future. In Huxley's view, only through scientific knowledge and wise decisions will human beings be successful on this planet or elsewhere.

In 1959, Julian Huxley wrote a favorable introduction to the English translation of Pierre Teilhard de Chardin's major work, *The Phenomenon of Man* (1938–1940).[3] As a geopaleontologist and Jesuit priest, Teilhard had attempted to reconcile the facts of science with the beliefs of Christianity in his own admittedly personal philosophy of evolution.[4] Not surprisingly, the theologian gave preference to Roman Catholicism and cosmic mysticism over science and reason.

As a secular humanist, Huxley could not accept Teilhard's leap of faith from scientific evolution to spiritual mysticism. Although a scientist, the Jesuit priest and religious humanist had envisioned a future end to the noosphere (the human world) as a final spiritual unity of humankind that will detach itself from the earth, transcend both space and time, and then immerse itself within a personal god at the Omega Point. By contrast, Huxley as secular naturalist maintained that the process of evolution has no purpose, direction, or ultimate goal. However, he did express a deep concern for the ongoing development of our species on this planet in light of those problems caused by the reckless use and wanton abuse of our earth.

Although Huxley never embraced Teilhard's theistic interpretation of evolution, which argued for cosmic teleology, pervasive spiritualism, and a mystical end-goal for our species, he was outwardly sympathetic to Teilhard's evolutionary synthesis, especially because it upheld evolution as a fact of reality. In addition, Teilhard and Huxley showed great concern for both scientific research and a need to improve the human condition within dynamic nature. Actually, Huxley himself had been advocating a comprehensive view of our species within an evolving universe.

In his rejection of Lysenkoism, the view of genetics in Russia early in this century, Huxley himself placed the scientific quest for truth above myopic politics and vested interests. In fact, Neodarwinism excludes all the explanatory mechanisms of Lamarckism that had pervaded Russian biology.

---

3. Pierre Teilhard de Chardin, *The Phenomenon of Man,* 2d ed. (New York: Harper Torchbooks, 1965).

4. Birx, *Interpreting Evolution,* pp. 178–222.

Huxley the optimist saw evolutionary humanism as a science of endless possibilities. For him, our species as the agent and spearhead of evolution on Earth is now responsible for the existence of life on this planet; it must draw up a long-term plan for its further evolution and future destiny in the universe. He welcomed the convergence of ideas and values, and advocated the one-world concept: a global view of humankind favoring science, freedom, planning, democracy, and tolerance. But the work of scientists and philosophers is never finished. Surely the cosmic perspective and the evolutionary framework must remain open to new facts, concepts, and hypotheses.

Scientific naturalism and secular humanism continue to challenge entrenched but outmoded beliefs, values, and worldviews. With confidence, Huxley maintained that evolutionary humanism will become our dominant idea-system of the future. It will allow for even more scientific achievement as well as the greater fufillment of human potentialities. According to Huxley, the ongoing challenge of science is to ascertain the place of humankind within this universe; as such, rational thought and science education are of major importance. He even took seriously the possibility that life exists on other worlds, thereby acknowledging the new science of exobiology.

Huxley's *Evolutionary Humanism* is a significant contribution to science-oriented thinking; its fourteen essays, which first appeared as a single volume in 1964, may still be read with great benefit. They exemplify Huxley's critical intellect while demonstrating the ongoing challenge of coming to grips with an evolving universe. These essays also extol the value of individual persons as well as the need for free and responsible inquiry. Finally, and quintessentially, they clearly testify to the urgent need for a secular humanist stance if our species is to survive and succeed free from fear, blind faith, ignorance, dogmatism, intolerance, and misplaced values.

Sir Julian Huxley died in London on February 14, 1975.

Huxley's other published works include: *The Individual in the Animal Kingdom* (1912), *Essays of a Biologist* (1923), *Religion without Revelation* (1927), *At the Zoo* (1936), *Toward a New Humanism* (1957), and *Memories* (1970–1973).

H. JAMES BIRX

Further Readings

Birx, H. James. *Interpreting Evolution: Darwin & Teilhard de Chardin*. Buffalo, N.Y.: Prometheus Books, 1991.

Clark, Ronald W. *The Huxleys*. New York: McGraw-Hill, 1968.

Dobzhansky, Theodosius. *Mankind Evolving: the Evolution of the Human Species*. New York: Bantam, 1970.

Haeckel, Ernst. *The Riddle of the Universe*. Buffalo, N.Y.: Prometheus Books, 1992.

Huxley, Julian S. *Evolution in Action* (1953). New York: Mentor Books, 1957.

————. *Evolution: The Modern Synthesis* (1942). New York: John Wiley & Sons, 1964.

————. *Religion Without Revelation* (1927). New York: Mentor Books, rev. ed., 1957.

Kurtz, Paul. *Eupraxophy: Living Without Religion*. Buffalo, N.Y.: Prometheus Books, 1989.

Simpson, George Gaylord. *This View of Life: The World of an Evolutionist*. New York: Harcourt, Brace & World, 1964.

Wilson, Edward O. *The Diversity of Life*. Cambridge: Belknap Press/ Harvard University Press, 1992.

CONTENTS

ACKNOWLEDGMENTS

MANY essays in this volume were published as separate lectures, as chapters in books, or articles in journals. I wish to make acknowledgment to the publishers for permission to reproduce them here. In particular, for *The Emergence of Darwinism* (from "Evolution after Darwin", ed. Sol Tax, 1960) to the Chicago University Press; for *Higher and Lower* (from the J. Roy. Coll. Surgeons, Edinburgh, 1962) to the Royal College of Surgeons of Edinburgh; for *Psychometabolism* to the Journal of Neuropsychiatry, Chicago; for *The Humanist Frame* (from the book of the same title) to Messrs. Allen and Unwin and Harper and Row; for *Education and Humanism* (9th Fawley Lecture, 1962) to the University of Southampton; for *Birds and Science* to Country Life Ltd; for *The Coto Doñana* to "Animals": for *Riches of Wild Africa* to the Observer; for *Toynbee and Time Scales* to the Sunday Times for *Teilhard de Chardin* (introduction to "The Phenomenon of Man" 1959) to Messrs. Collins and Harper and Row; for *The New Divinity* to The Twentieth Century; for *The Enlightenment and the Population Problem* (from Transactions of the First International Congress on the Enlightenment II, ed. T. Besterman) to Institut et Musée Voltaire, Les Délices, Geneva; for *The Crowded World* to Punch and to the Eugenics Review; and for *Eugenics in Evolutionary Perspective* (the Galton Lecture, 1962) to the Eugenics Review.

THE EMERGENCE OF DARWINISM

IN 1958 we celebrated the centenary of an outstanding event in the history of science—the birth of Darwinism or evolutionary biology, initiated by the joint contribution of Charles Darwin and Alfred Russel Wallace to The Linnean Society of London, announcing their independent discovery of the principle of natural selection.

I say Darwinism because not only did Darwin have priority in conceiving that evolution must have occurred, and could only have occurred through the mechanism of natural selection, but he also contributed far more than Wallace, or indeed than any other man, to the solution of the problem and the development of the subject. I shall therefore speak almost entirely about Darwin and Darwinism, endeavouring to bring out facts and ideas which illuminate Darwin's unique role in the history of our science.

Charles Darwin has rightly been described as the "Newton of biology": he did more than any single individual before or since to change man's attitude to the phenomena of life and to provide a coherent scientific framework of ideas for biology, in place of an approach in large part compounded of hearsay, myth, and superstition. He rendered evolution inescapable as a fact, comprehensible as a process, all-embracing as a concept.

His industry was prodigious. His published books run to over 8000 printed pages and contain, on my rough estimate, at least 3,000,000 words. His scientific correspondence must have reached similar dimensions, and his contributions to scientific journals comprise well over 400 pages.

The range of subjects with which he dealt, often as an initiator and always magisterially, was equally remarkable. Let us first recall that at the outset of his career he was more of a geologist than a biologist, that his first scientific works, on coral reefs and on the geology of South America, dealt with geological subjects, and that the only professional

9

position he ever occupied was that of Secretary to the Geological Society. Later, he dealt with the taxonomy and biology of that "difficult" group of animals, the barnacles or Cirripedes, in its entirety; with the principles and practice of classification; with the evidences for evolution; the theories of natural and sexual selection and their implications; the descent of man, including the evolution of his intellectual, moral, and aesthetic faculties; the emotions and their expression in men and animals; geographical distribution, domestication, variation in nature and under domestication; the effects of self- and cross-fertilization (or, as we should now say, in- and out-breeding) and various remarkable adaptations for securing cross-fertilization; the movements of plants, insectivorous plants, and the activities of earthworms.

Not only is he the acknowledged parent of evolutionary biology, but he is also prominent among the founding fathers of the sciences we now call ecology and ethology.

Above all, he was a great naturalist, in the proper sense that he was profoundly interested in observing and attempting to comprehend the phenomena of nature, though at the same time he managed to keep abreast of pure scientific advance in the fields which concerned him, such as general botany, embryology, paleontology, biogeography, taxonomy, and comparative anatomy, as well as with the activities both of professionals and amateurs in what we should now call plant and animal breeding.

He had an inborn passion for natural history, which showed itself from early childhood. Later, like most true naturalists, besides being motivated by intellectual interest, he was deeply moved by the wonder and beauty of nature. As a young man, he found an "exquisite delight in fine scenery", and enjoyed exploring wild and strange country.

The combination of deep emotion with close observation appears vividly in his notes on his first experience of the tropical rain-forest: "Twiners entwining twiners—tresses like hair—beautiful lepidoptera—Silence—hosannah—frog habits like toad—slow jumps." "Sublime devotion the prevalent feeling." And a little later, "Silence well exemplified. ...Lofty trees, white boles. ... So gloomy that only shean [*sic*] of light enters the profound. Tops of the trees enlumined."

I may perhaps note that this last entry was made, not in the remote depths of the great Amazonian forest as one might expect, but close to Rio, at Botofogo, whose beach is now bordered by luxury hotels and crowded with bathing beauties. However, though roads have robbed the forest behind the beach of its primal virginity so that one might now call it *demi-vierge*, it is otherwise untouched, and in its recesses one can still recapture some of Darwin's feelings.

Another characteristic of Darwin was his extraordinary diffidence, coupled with a passion for completeness and a reluctance, so extreme as to appear almost pathological, to publish to the world his ideas on the controversial subject of evolution before he had buttressed his arguments with a body of evidence which would overwhelm opposition by its sheer vastness. It has been suggested that these traits in Darwin's character, and also the constant ill-health from which he suffered after his marriage in 1839, were neurotic symptoms springing from unconscious conflict or emotional tension, and that this in its turn was first generated by Darwin's ambivalent attitude to the dominating and domineering figure of his father, Robert Darwin.

Although a contributory cause seems definitely to have been a trypansome infection (Chagas's Disease) caught during the voyage, his ill-health was assuredly in part an escape mechanism, fostered by his wife's devotion, who became the ideal sick-nurse as Darwin became the ideal patient. His reluctance to commit himself publicly and in print to belief in the mutability of species and in evolution by natural causes sprang ultimately from some unacknowledged inner conflict which was partly rooted in his relations with his father. It was his father who took him away from school early because he thought he was idle and doing no good; who decided first that he should study medicine, and then, when it was clear that Charles disliked the prospect of becoming a physician, sent him to Cambridge to study for the Church, another profession for which he had no inclination or aptitude; and whose strong opposition to Charles accepting the post of naturalist on the *Beagle* nearly robbed the world of its greatest biologist. He clearly deplored Charles' intense devotion to nature and natural history, which was manifested in the pursuits of his childhood and youth, from beetle-collecting to shooting and geologizing in the field. Furthermore, his father was a man of decided opinions, very autocratic with his children, and apparently hostile to the whole idea of evolution. In his autobiography Charles states that he never heard the idea of evolution favourably mentioned until he had gone as a medical student to Edinburgh: this at least indicates that it was not discussed in the Darwin home. In any case, what could be more symptomatic of a guilt complex than Darwin's confession, in a letter to Hooker early in 1844, that to assert that species are not immutable is "like confessing a murder"! If he felt like this, it is little wonder that he kept on putting off the public statement of his views.

Furthermore, the conflict must have been sharpened by his marriage, for his deeply religious wife was opposed to all unorthodox views. In any case, his chronic ill-health did not begin until after his marriage.

His extreme diffidence about the merits of his work (clearly another symptom of inner conflict) is illustrated by a letter of 27 August 1859 to his publisher, John Murray, about the "little work"—as he called the *Origin of Species*—which he was then preparing. "I feel bound [he wrote] for your sake and my own to say in clearest terms that if after looking over part of my MS. you do not think it likely to have a remunerative sale I completely and explicitly free you from your offer."

It is worth while retelling the salient facts of the story. During the voyage of the *Beagle*, probably towards the end of 1835, he had become convinced that species could not be separate immutable creations. In 1837, soon after his return to England, he started a series of notebooks on the "transmutation of species", in the full consciousness that this would imply large-scale evolution and the common ancestry of all organisms, including man. He soon realized the efficacy of selection in creating new varieties of races of domestic animals and plants, but was unable to see how it could operate in nature. Then, late in 1838 he "happened to read for amusement *Malthus* on *Population*"—I quote his own revealing phrase in his Autobiography—and the idea of natural selection immediately flashed upon him. "Here then," he continued, "I had at last got a theory to which to work." This vivified all his subsequent thinking: for do not let us forget that Darwin combined inductive and deductive method in a remarkable way. He was never interested in facts for their own sake, but only in their relevance to some hypothesis or general principle. But when he had discovered some satisfactory general principle, he proceeded to deduce the most far-reaching conclusions from it. This is particularly evident, as will appear later, with the principle of natural selection; but it is also true of his treatment of uniformitarianism and the principles of continuity, of sexual selection, and of biological adaptation.

This is perhaps the place to stress another aspect of Darwin's mind. Although his laborious patience in the collection and synthesis of factual evidence has rarely been rivalled (he himself called his mind "a kind of machine for grinding general laws out of large collections of facts"), yet sudden intuition was responsible for some of his most important discoveries of principle, notably natural selection and the explanation of biological divergence—a valuable reminder of the fact that imagination as well as hard work is essential for scientific comprehension.

But I must return to my story. In spite of this illuminating discovery, his reluctance to commit himself was such that not until four years later did he "allow himself the satisfaction" (again a revealing phrase) of putting his ideas on paper; and then only by "writing out in pencil a very brief abstract" of his theory and the evidence for it.

Two years later, in 1844, he enlarged this into an "Essay". As a matter of fact, this so-called essay was a sizeable book of 230 pages, covering almost the same ground as the *Origin*, and more than adequate as an exposition of the whole subject. Yet he still procrastinated, and continued to procrastinate for fourteen further years. He showed the essay to no one but Lyell and discussed his evolutionary ideas only with him and a few intimate colleagues, notably Hooker. He continued with the interminable collection of facts, until finally, urged on by Lyell and Hooker, he began in 1856 to write a monumental work on the subject.

Here I must pay tribute to Alfred Russel Wallace. I wish I had more space to set forth his great contribution to evolutionary biology. He laid the foundations of zoogeography, and his notable works on the subject—*the Geographical Distribution of Animals* and *Island Life*—can still be read with profit, as can those on tropical natural history in general —*Tropical Nature* and *The Malay Archipelago*. He was the first to make a comprehensive analysis of cryptic adaptations; he contributed materially to the study of mimicry, and originated the theory of warning coloration. He made many original contributions to the species problem, and in 1855 had published a paper, "On the Law which has regulated the Introduction of New Species" (*Ann. May. Nat. Hist.*, 1855, p. 184), which showed that he believed in the evolution of new species from old, and led to Darwin entering into correspondence with him.

But not only was he a great naturalist, not only did he independently discover the principle of natural selection, but by doing so he forced Darwin into publication. If it had not been for Wallace's attack of malarial fever in Ternate and his impulsive temperament, the *Origin of Species* would never have been published in 1859. Ever since 1855, when he had become convinced that evolution had occurred, the question of *how* changes of species could be brought about was constantly in his thoughts, but he never succeeded in thinking the problem out. The fever, by setting him free from his daily routine of practical detail, permitted his roving mind to discover the principle of natural selection (as with Darwin, in a sudden flash of intuition, and also as a result of reading Malthus and Lyell some time previously); and his temperament, the very opposite of Darwin's, led him to write down his ideas that same evening, to elaborate them during the next two days, and then send them straight off to Darwin for his opinion. The first result, after much heart-searching on Darwin's part and the firm intervention of

14

Lyell and Hooker, was the joint announcement of Darwin's and Wallace's view to the Linnean Society of London on 1 July 1858, and their subsequent publication in the Society's Journal. The second and much more important result was the publication of the *Origin of Species*. Strongly pressed by Lyell and Hooker, in September 1858, Darwin started "abstracting" (his own word) his huge incomplete work, and finished the book in just over thirteen months. Although in his autobiography he still called it "only an abstract", he acknowledged that it was "no doubt the chief work of my life", and this is certainly true.

But for Wallace and his fever, Darwin would assuredly not have overcome his resistance to speedy publication, and would have continued working on "the MS. begun on a much larger scale". In 1858 he envisaged its completion "at the soonest" by 1860. But we can be sure that his inhibitions over coming into the open, which were transmuted into perfectionist dreams of completeness ("I mean to make my book as perfect as ever I can," he wrote as late as February 1858), would have prevented him from publishing for a much longer time—perhaps five, perhaps even ten years.

He himself said that the book would have been "four or five times as large as the *Origin*"—which would mean at least 2500 pages, and over three-quarters of a million words!—and that very few would have had the patience to read it. It would, indeed, have been almost unreadable, and the forceful flow of argument, so well manifested in the *Origin*, would have been lost in the sands of over-abundant fact. Biology certainly owes a great deal to Wallace.

Nor must we forget Lyell. He was the chief source of encouragement to Darwin in his evolutionary work after his return to England and was mainly instrumental in persuading him to publish his ideas together with Wallace's paper in 1858. We know that his *Principles of Geology* influenced Wallace more than any other book. Above all, his great work demonstrating that slow geological change had occurred as a result of existing physical causes prepared the ground for the idea of biological evolution by natural means. As T. H.

Huxley wrote in 1887, he was "the chief agent in smoothing the path for Darwin".

Biology also owes a good deal to Darwin's caution, exaggerated though this was. If Darwin had rushed into print in 1838 with a brief and bare account of his conclusions, they would have been stillborn. The idea of evolution needed heavy reinforcement with facts, and the idea of natural selection had to be thoroughly worked out in all its implications. Even though the *Essay* of 1844 went a long way towards satisfying these requirements, its immediate publication would not, I am sure, have been nearly so effective as was that of the *Origin* fifteen years later. This is partly owing to Darwin's enlargement of his evidence and improvement of his argument, but also to the current of opinion, the increased interest of biologists in evolution and their increasing readiness to discuss it, as well as to the appearance on the biological stage of younger men, like Wallace, Alfred Newton, and especially Huxley, ready to be persuaded and become forceful champions of the new and revolutionary ideas. The best time for Darwin to publish was, I would say, between 1855 and 1860.

Above all, delay in publication gave Darwin time to look at every aspect of his enormous subject, to think out its many implications, and to meet all possible objections. The result was extremely impressive, and far more convincing than any brief sketch, however brilliant, or any speculative picture, such as those drawn by Erasmus Darwin and by Lamarck.

The last paragraph of the *Origin* has often been quoted: I quote it here once again, as admirably illustrating this close-reasoned comprehensiveness of Darwin's work.

It is interesting to contemplate a tangled bank, clothed with many plants of many kinds, with birds singing on the bushes, with various insects flitting about, and with worms crawling through the damp earth, and to reflect that these elaborately constructed forms, so different from each other, and dependent upon each other in so complex a manner, have all been produced by laws acting around us. These laws, taken in the largest sense, being Growth with Reproduction; Inheritance which is almost implied by reproduction;

Variability from the indirect and direct action of the conditions of life, and from use and disuse; a ratio of increase so high as to lead to a Struggle for Life, and as a consequence to Natural Selection, entailing Divergence of Character and the Extinction of less-improved forms. Thus, from the war of nature, from famine and death, the most exalted object which we are capable of conceiving, namely, the production of the higher animals, directly follows. There is grandeur in this view of life, . . . that, whilst this planet has gone cycling on according to the fixed laws of gravity, from so simple a beginning endless forms most beautiful and most wonderful have been, and are, being evolved.

It is interesting to pursue the question of timing on to a more speculative plane, and ask ourselves what would have happened to Darwin if he had been born a century earlier or a century later. I would guess that if he had been born in 1709 he might well have become a good amateur naturalist, rather after the pattern of his grandfather Erasmus, one who would, perhaps, have indulged in some interesting specula-tions on natural history, but would not have been likely to make any major discoveries or to exert any important influence on scientific or general thinking. If he had been born in 1909 he might at most, I would hazard, have achieved some eminence as a professional ecologist. In the one case the time was unripe, in the other over-ripe.

Kroeber has demonstrated that the effective manifestation of genius requires not only exceptional individual talent, but depends also on the circumstances and sometimes the accidents of place and period; nowhere is this better illustrated than in the person of Darwin. First of all, the scientific and intellectual atmosphere was propitious. The time was just ripe for the tying together of the facts of geology and biology by the unifying principles of evolution. Then, as a boy and young man, Darwin was able to indulge his early taste for natural history; later, his financial in-dependence enabled him to devote himself entirely to his own chosen work, and his invalidism prevented him from wasting time and energy in a round of social engagements and scientific meetings; as an Englishman, he quickly came into contact with the ideas of Malthus, Lyell, and Hooker, which were so decisive for his thought, and with Huxley, who was so important in spreading his doctrine; above all, he had the luck to go as naturalist on the *Beagle*.

Two circumstances of the voyage seem to have been of outstanding importance. First, he was able to study natural history, in its geological as well as in its biological aspects, on a continental scale, and so to appreciate the overall pattern of the fauna and flora, and also their gradual transitions and modifications of detail in relation to changing conditions of time and place. This forced him to think along broad lines, in terms of continuity and gradual evolutionary change: in a way that would hardly have been possible if he had stayed at home. In a similar fashion the small extent but great geological variety of Britain prevented its scientists grasping the general principles of soil science, while the great expanses and broad zonation of the Russian landscape facilitated or even forced their recognition by Soviet pedologists.

The other decisive circumstance was the *Beagle*'s visit to the oceanic archipelago of the Galapagos. Oceanic archipelagos are rare natural laboratories, in which enquiring and receptive minds can find a demonstration of evolution and how it operates in practice. Darwin's mind was both enquiring and receptive: it seems clear that his experiences here finally crystallized his thought and convinced him that evolution was a fact.

Here biology must acknowledge its very real debt to Darwin's uncle, Josiah Wedgwood. Robert Darwin's objections to Charles' accepting the post of naturalist on the *Beagle* were so strong (and his influence on his son so powerful) that Charles, though eager to accept, wrote to refuse the offer. And it was only his uncle's intervention that persuaded his father to withdraw his objections.

Robert Darwin seems to have taken a rather poor view of Charles' abilities and character. In fact, however, these must already have been impressive at the age of 22. They impressed Henslow and Sedgwick at Cambridge; the Hydrographer to the Navy, in a letter to Captain Fitzroy, his future commanding officer, speaks of him as "full of zeal and enterprise and having contemplated a voyage on his own account to South America"; and Captain Fitzroy himself wrote to the Hydrographer on 15 August 1832, that "Mr Darwin is a very superior young man, and the very best that could have been detailed for the task".

But I must return to my central theme. Whatever the contribution of others, Darwin stands out as the prime author and pre-eminent figure of the biological revolution. Wallace himself fully recognized this. It was he who first called Darwin "the Newton of Natural History" (or biology, as we should say today), and he coined the term *Darwinism* as the title of his own book on evolution. The evidence and the arguments marshalled by Darwin in the *Origin* were decisive in persuading leaders of scientific thought like Huxley and Hooker that evolution had occurred and that it was based on a natural and scientifically intelligible mechanism.

Furthermore, his inhibitions over publication disappeared with the appearance of the *Origin*, and he proceeded to develop various aspects of the subject with remarkable speed and energy. Twenty-two years elapsed between his opening his notebooks on the transmutation of species and the publication of the *Origin*, and fourteen years between the writing of the *Essay* and the appearance of the joint paper with Wallace. In the fourteen years after 1859 he published three truly major works—*The Variation of Animals and Plants under Domestication*, *The Descent of Man and Selection in Relation to Sex*, and *The Expression of the Emotions in Man and Animals*—and two minor (though still important) ones; and if we take the period of twenty-two years we have to add five more volumes, ending with his last book, the fascinating study of earthworms.

The emergence of Darwinism, I would say, covered the fourteen-year period from 1858 to 1872; and it was in full flower until the 1890's, when Bateson initiated the anti-Darwinian reaction. This in turn lasted for about a quarter of a century, to be succeeded by the present phase of Neo-Darwinism, in which the central Darwinian concept of natural selection has been successfully related to the facts and principles of modern genetics, ecology, and palaeontology.

When we biologists take stock of our subject today, we speedily discover the magnitude of Darwin's contribution; we see how much of his thought has become incorporated in the permanent framework of our science, how many of his ideas are still alive and fruitful. In the first place, we build on his demonstration that evolution has taken place, and has taken place by natural means, so that both its course and its mechanism can be further investigated by scientific methods. Then his ideas of continuity and gradual transformation remain basic for evolutionary biology—abrupt changes of large extent, as in polyploidy, are exceptional. He stressed the importance of time as a factor in evolution: for selection to produce changes of large extent, time must be forthcoming in enormous quantities—how enormous, we have only recently realized. It is by following out such ideas that evolutionary biologists are now calculating the actual rates of evolution in different groups.

The principle of natural selection was Darwin's greatest discovery, and it remains central to all biological thinking. Darwin's tenacious and comprehensive mind insisted on deducing all possible general conclusions from the principle and on pursuing its implications to the limit. Thus natural selection, he saw, implied that evolutionary change would be gradual and slow. But perhaps his conclusions on biological improvement afford the most remarkable example of his capacity for bold yet careful generalization. Natural selection, he wrote, has as its "ultimate result . . . that each creature tends to become more and more improved in relation to their conditions. This improvement inevitably leads to the gradual advancement of the organisation of the greater number of living beings throughout the world."

The first sentence refers to small-scale processes and makes intelligible the omnipresence of detailed adaptation, or biological fitness, as some modern workers prefer to call it. It also implies the point made explicitly by Darwin elsewhere that natural selection can never produce characters which are solely or primarily useful to another species. The second sentence, referring to long-term evolution, extends the idea of improvement to cover improvement in general

20

organization, and seems to be the first scientifically based argument for the inevitability of some degree of biological progress or evolutionary advance.

He saw the implications of intra-sexual competitive selection in producing masculine weapons, and of inter-sexual allaesthetic selection in generating masculine adornments and displays. In *The Expression of the Emotions* he laid the foundations for the modern science of comparative ethology. The very title of the book illustrates his robust naturalism: he saw clearly that the mental and physical characters of organisms are inseparable, and that emotions and intelligence must evolve as much as brains and bodily organization. He did not hesitate to extend his argument to cover man's distinctive mental capacities, intellectual, aesthetic, and moral. While subscribing to the view that "the moral sense or conscience constitutes by far the most important difference between man and lower animals", he considered that it had evolved naturally. I cannot forbear from quoting one characteristic passage:

> The following proposition seems to me in a high degree probable —namely that any animal whatever, endowed with well-marked social instincts, the parental and filial affections being here included, would inevitably acquire a moral sense or conscience, as soon as its intellectual powers had become as well, or nearly as well developed, as in man

(though, he adds, it might not be identical with ours). And later he states that the belief in spiritual agencies naturally follows from other mental powers.

It is clear that Darwin had fully grasped the important point that certain characters are what may be called consequential, arising in evolution as a consequence of the prior appearance of some other character, or because correlated with a change brought about by natural selection. Elsewhere Darwin stated this conclusion in general terms: "Owing to the Laws of Correlation, when one part varies or the variations are accumulated through natural selection, other variations, often of the most unexpected nature, will ensue."

Another of Darwin's notable deductive conclusions concerns divergence (or *cladogenesis*, as Rensch has called it). He was the first to realize that natural selection will lead inevitably to evolutionary divergence, both the small-scale divergence of related species, and the large-scale divergence which results in the appearance of distinct and well-defined group-units—genera, families, orders—in a hierarchical arrangement. Through the process of divergence each species exploits the resources of the environment more effectively, so that the large-scale result of divergence in the inhabitants of a region is comparable to the physiological division of labour in an individual body.

Darwin was the first to see the evolutionary explanation of the facts, later incorrectly subsumed by Haeckel under the head of an inevitable recapitulation of evolutionary development or phylogeny by individual development or ontogeny, concerning "the wide difference in many classes between the embryo and the adult animal, and of the close resemblance of the embryos within the same class".

His studies on cross-fertilization, and the mechanisms for securing it, paved the way for modern work on heterosis or hybrid vigour (and its application in the hybrid corn industry), and for a general theory of breeding systems, such as C. D. Darlington has so successfully propounded. In combination with his exhaustive survey of variation under domestication, they contributed materially to the development of the sciences of plant and animal breeding.

Finally, I must mention his conclusions on the processes by which new and successful types originate. While recognizing the importance of isolation, which we now regard as a necessary prerequisite for the separation of one species into two, he laid greater stress on the numerical abundance of the evolving species and the size of the area occupied by it. Greater abundance gives more chance for favourable variations to occur; greater size and diversity of area leads to more vigorous competition for survival, as well as providing greater opportunities for temporary isolation. All this will promote more rapid evolution, and the successful types will have a greater capacity for dispersal and for further evolutionary differentiation. In this, Darwin anticipated to a con-

22

siderable extent modern views on the factors underlying the origin, spread, and diversification of new types, new unit-steps in the evolutionary process.

It is also, I think, of interest to examine some of Darwin's errors and omissions in the light of our present knowledge. His theory of sexual selection has been the target for bitter and sometimes violent attack. It is true that he did lump together various kinds of display, notably hostile display against rivals and sexual display to potential mates; and that he ascribed much too great importance to female choice. But he grasped the essential point that striking displays must have a biological significance and must be what we now call *allaesthetic* in character, exerting their effect by stimulating the emotions of another individual via its visual or auditory senses. And he was quite correct in ascribing the evolution of masculine weapons to intra-sexual selection as between competing males.

Strangely enough, though he mentions cases where adornments are equally developed in both sexes, he dismissed the possibility of biologically effective mutual display between the actual or potential mates. Yet such displays are frequent and often striking, and must have been seen by naturalists before Darwin wrote the *Descent of Man*. I suspect that he was too deeply committed in his thinking to the ideas of female choice and male competition to envisage the possibility of mutual allaesthetic stimulation. Further, in his treatment of the subject he states that sexual selection "acts in a less rigorous manner than natural selection", because "the latter produces its effects by the life or death at all ages of the more or less successful competitor", while with the former, the less successful males merely "leave fewer, less vigorous or no offspring".

This strange error springs, I would guess, from his failure —perhaps inevitable at the time—to think quantitatively on the subject, coupled with his adoption of the phrase *the struggle for existence*, with its implications of an all-or-nothing competition, life or death. If he had ever spelled out natural selection in modern terms, as being the result of the differential reproduction of variants, he would at once have seen that any form of selection can vary in rigour according to circumstances, and indeed that intra-sexual selection between males in a polygamous species is likely to attain maximum selective intensity.

Strangely enough, elsewhere Darwin drops his all-or-nothing view and assumes a differential action of natural selection. This is, so far as I know, the one major point which he failed to think out fully and on which he expressed divergent conclusions at different times.

Though Darwin, like T. H. Huxley, thought very little of Lamarck's views on the mechanism of evolution—in a letter of 1844 to Hooker he writes "Heaven defend me from Lamarck nonsense of a 'tendency to progression', 'adaptation from the slow willing of animals, etc.' "—he did believe in the inheritance of certain "acquired characters"—the effects of the conditions of life and of use and disuse. Furthermore, he attached more importance to them in later editions of the *Origin*. It is this error, which for want of a better term we may loosely call Lamarckian, with which present-day biologists most often reproach Darwin.

It must be stressed, however, that he regarded these agencies as quite subsidiary to natural selection, which he consistently maintained was much the most important agency of evolutionary change.

These "Lamarckian" errors clearly sprang from the total ignorance of nineteenth-century biology on the subject of heredity. Fleeming Jenkin pointed out in 1867 that, on the current theory of blending inheritance, even favourable new variations would tend to be swamped out of effective existence by crossing, if heritable variation in general was rare and infrequent. It was to provide for sources of more abundant variation that Darwin came to ascribe increased importance to the evolutionary role of "acquired characters". Only when the actual genetic mechanism had been discovered and its particulate (non-blending) nature had been established, could it be shown—notably by R. A. Fisher— that Lamarckian (and orthogenetic) theories of evolution were not only unnecessary but inherently incorrect and impossible.

Disuse often does result in evolutionary degeneration: but,

as H. J. Muller has shown, this is the result of mutation and selection, not of the direct inheritance of somatic effects.

Changed conditions again may have evolutionary results —but once more, not through their direct effects. They may result in increased variability, as Darwin stressed. But this is merely due to rare mutants and new combinations being able to survive in the altered conditions and also to their arising as a result of inbreeding.

In other cases a character which looks like a modification, a direct response to environmental conditions, turns out to be hereditary. We now know that such apparently Lamarckian results may be obtained in a non-Lamarckian way, by what Waddington calls *genetic assimilation*. With characters which in normal stock are only produced by special environmental stimuli, selection of those individuals showing the character in extreme form may, in a comparatively few generations, lead to the character appearing in a few individuals without exposure to the special stimulus; and further selection, in normal environmental conditions, will produce an overwhelming majority permanently showing the character.

The developmental process leading to the phenotypic manifestation of such a character has both environmental and genetic determinants. During assimilation the genetic determinant has been strengthened, by selection for genes favouring manifestation, to a point at which the process has been genetically canalized and the environmental determinant is no longer required. But since selection acts not on genotypes but on phenotypes, the environmental determinant was originally necessary to produce something on which selection could operate. The result is a modernized version of Baldwin and Lloyd Morgan's organic selection. Thus assimilation, not the inheritance of acquired characters in the usual sense, could account for the origin of various adaptations, such as genetically determined callosities in the exact situations where they are specially required, as in ostriches and warthogs, and many adaptive features of plant ecotypes.

Other adaptations, however, such as those of the hard

25

parts of holometabolous insects, or those involving mimetic resemblance, demand explanation (as Darwin fully realized) in terms of natural selection acting on adaptively random genetic variation. But when virtually nothing was known about the mechanism of reproduction, heredity and development, many phenomena were more readily interpreted on a non-selectionist basis.

It has been suggested that Darwin would have avoided falling into these pitfalls if only he had paid attention to Mendel's work, which was published in 1865, in plenty of time for Darwin to amend his views in later editions of the *Origin*. I do not think this is so. It needed nearly twenty years of intensive research on suitable material such as *Drosophila* before the findings of genetics could be fruitfully integrated with evolutionary theory. Before that, most geneticists, obsessed by the obvious mutations with large effects which they naturally first studied, were led to anti-selectionist views and to the idea that evolution would normally take place by discontinuous steps, or even merely as the result of mutation-pressure. Only when they had arrived at a true picture of the genetic constitution as a flexible gene-complex in which many genes of small effect collaborate to produce phenotypic characters, only then could they see that discontinuity in the genetic basis of variation need not imply discontinuity in its phenotypic manifestation. Consequently evolutionary change, though due to selection of genetically discontinuous variants, can normally be continuous.

Darwin had already arrived at this correct conclusion without any knowledge of the underlying mechanisms involved. With his usual common sense he concentrated on phenotypes; accordingly, continuous variation and gradual change became essential in his thought. I suspect that if he had known of Mendel's results he would have regarded them as interesting but exceptional and relatively unimportant for evolution, as he had already done for other cases of large mutations and sharp segregation. A premature attempt at generalizing Mendelian principles would merely have weakened the central Darwinian principle of gradual slow change.

26

There is, finally, Darwin's failure to recognize explicitly the radical differences between man and other animals, especially between the process of evolution in man and in other animals. It is true that he speaks of high intellectual power and conscious morality as distinctive attributes of our species and implies that human speech is something *sui generis* as a means of communication; it is true that he regards man as the highest product of evolution. But nowhere does he point out man's truly unique and most important characteristic—cumulative tradition, the capacity for transmitting experience and the fruits of experience from one generation to the next; nor does he discuss the implications of this new human mechanism of change, as he did so exhaustively for the biological mechanism of natural selection. Thus, while overwhelmed by the thought that modern Europeans must be descended from ignorant savages, like the naked Fuegians who burst on his astonished sight, he makes no attempt to discuss or even to point out the fact that evolution from the savage to the civilized state involves essentially not a biological but a cultural change.

Why was this? I suggest that it was because Darwin's primary and main aim was to provide convincing evidence that organisms were not immutable creations but had evolved by natural means from something different; and this implied a focusing of attention on their past history. This preoccupation of his with origins is revealed in the titles he chose for his two greatest works—the *Origin of Species* and *The Descent of Man*—though *The Evolution of Organisms* and *The Ascent of Man* would in fact have been more appropriate.

His tactics were probably sound: at the time, the main need was to establish on a firm basis the *fact* of evolution and its scientific comprehensibility. In recent years, however, we have turned our attention to the *course* of evolution; and as a result, have been enabled to reach a number of important conclusions about the evolutionary process in general, and our own place and role within it in particular. This has been largely thanks to the soundness of the foundations, both of fact and of idea, provided by Darwin.

That evolution is a natural process, involving man as well

27

as all other organisms in its unbroken continuity: that natural selection inevitably generates novelty, adaptive improvement, and advance in general organization: that successful types tend to differentiate into dominant groups: that improvement of the mental capacities of life, or as I would prefer to put it, advance in the organization of awareness, has been one of the most striking trends in the evolution of higher animals, and has led naturally to the appearance of the distinctive mental and moral qualities of man—these ideas of Darwin, I would say, have been especially important for the later development of evolutionary theory.

The study of evolution's course, following up Darwin's ideas on divergence and the formation of dominant groups, has revealed that evolutionary advance occurs in a series of steps, through a succession of dominant types. This is the result of very long-term selection, selection between types or groups instead of between individuals. The more efficient type will automatically tend to spread and differentiate at the expense of the less efficient: it is as simple as that. As a result, the more efficient type evolves into a large and successful group, while earlier groups with which it competes are reduced. Taxonomic groups are thus organizational grades as well as phylogenetic units. And the grade is the unit-step of evolutionary advance. On the large and long-term scale this process results in the familiar but essential fact of the succession and replacement of large dominant groups, each embodying some important new improvement and constituting a new organizational grade.

Sooner or later, each group realizes all its inherent possibilities and becomes stabilized, incapable of major advance except through the rare event of some line evolving an organization with new advantages, and so permitting a breakthrough to a new grade of advance. This, it seems, can never happen twice, for competition with the established successful type will automatically prevent a second invasion of the same evolutionary territory or *adaptive zone*, as modern evolutionists like G. G. Simpson call it.

This was an important clarification of the biological scene. Meanwhile, the window that Darwin opened into the world

28

THE EMERGENCE OF DARWINISM

of life permitted a new and evolutionary view of other subjects. Men began studying the evolution of nebulae and stars, of languages and tools, of chemical elements, of social organizations. Eventually they were driven to view the universe at large *sub specie evolutionis*, and so to generalize the evolutionary concept in fullest measure. This extension of Darwin's central idea—of evolution by natural means—is giving us a new vision of the cosmos and of our human destiny.

Evolution in the most general terms is a natural process of irreversible change, which generates novelty, variety, and increase of organization: and all reality can be regarded in one aspect as evolution. Biological evolution is only one sector or phase of this total process. There is also the inorganic or cosmic sector and the psychosocial or human sector. The phases succeed each other in time, the later being based on and evolving out of the earlier. The inorganic phase is pre-biological, the human is post-biological. Each sector or phase has its own characteristic method of operation, proceeds at its own tempo, possesses its own possibilities and limitations, and produces its own characteristic results, though the later phases incorporate some of the methods and results of the earlier ones.

The cosmic phase operates by random interaction, primarily physical but to a small degree chemical. Its quantitative scale is unbelievably vast both in space and time Its visible dimensions exceed 1000 million light-years (10^{22} km), its distances are measured by units of thousands of light-years (nearly 10^{16} km), the numbers of its visible galaxies exceed 100 million (10^8) and those of its stars run into thousands of millions of millions (10^{15}). It has operated in its present form for at least 6000 million years, possibly much longer. Its tempo of major change is unbelievably slow, to be measured by 1000-million-year periods. According to the physicists, its overall trend, in accord with the Second Law of Thermodynamics, is entropic, tending towards a decrease in organization and to ultimate frozen immobility; and its products reach only a very low level of organization—photons, subatomic particles, atoms, and simple inorganic

29

compounds at one end of its size-scale, nebulae, stars and occasional planetary systems at the other.

On our earth and probably on a number of other planets, conditions favoured the production of more complex chemical compounds, culminating in substances, capable of self-reproduction and self-variation, and therefore subject to a new mechanism of change—natural selection. The passing of this critical point initiated the biological or organic phase of evolution, which proceeded at a much quicker tempo, produced far more variety, and reached far higher levels of organization. Its phase operates primarily by the teleonomic or ordering process of natural selection, which canalizes random variation into non-random directions of change. Its tempo of major change is somewhat less slow than that of the cosmic phase, measured by 100-million- instead of 1000-million-year units of time. Its overall trend, kept going of course by solar energy, is anti-entropic, towards an increase in the amount and quality of adaptive organization. And its results are organisms—organisms of an astonishing efficiency, complexity, and variety, almost inconceivably so until one recalls R. A. Fisher's profound paradox, that natural selection plus time is a mechanism for generating an exceedingly high degree of improbability.

In the course of biological evolution, three sub-processes are at work. The first (cladogenesis, or branching evolution) leads to divergence and greater variety; the second (ana-genesis, or upward evolution) leads to improvement of all sorts, from detailed adaptations to specializations, from the greater efficiency of some major function like digestion to overall advance in general organization; the third is stasi-genesis or stabilized limitation of evolution. This occurs when specialization for a particular way of life reaches a dead end as with horses, or efficiency of function attains a maximum as with hawks' vision, or an ancient type of organization persists as a living fossil like the lung-fish or the tuatara. The great novelty of the biological phase was the emergence of awareness—psychological or mental capacities—to a position of increasing biological importance.

Eventually, in the line leading to man, the organization of

awareness reached a level at which experience could be not only stored in the individual but transmitted cumulatively to later generations. This second critical point initiated the human or psychosocial phase of evolution. Spatially this phase is very limited; we know of it only on this earth, and in any case it must be restricted to the surface of a small minority of planets in the small minority of stars possessing planetary systems. On our planet it is at the very beginning of its course, having begun less than one million years ago. However, its tempo is not only much faster than that of biological evolution, but manifests a new phenomenon, in the shape of a marked acceleration. Its overall trend is highly anti-entropic, and is characterized by a sharp increase in the operative significance of exceptional individuals and of true purpose and conscious evaluation based on reason and imagination as against the automatic differential elimination of random variants. In this phase, though natural selection and physico-chemical interaction continue to operate, they are subsidiary to the new mechanism of change based on cumulative cultural tradition and especially on the growth and improved organization of tested knowledge. And its results are psychologically (mentally) generated organizations even more astonishingly varied and complex than biological organisms—machines, concepts, cooking, mass communications, cities, philosophies, superstitions, propaganda, armies and navies, personalities, legal systems, works of art, political and economic systems, entertainments, slavery, scientific theories, hospitals, moral codes, prisons, myths, languages, torture, games and sports, religions, record and history, poetry, civil services, marriage systems, initiation rituals, agriculture, drama, social hierarchies, schools and universities.

In the psychosocial phase of evolution the same three subprocesses operating in the biological phase are still at work—cladogenesis, operating to generate difference and variety within and between cultures; anagenesis, operating to produce improvement in detailed technological methods, in economic and political machinery, in administrative and educational systems, in scientific thinking and creative expression, in moral tone and religious attitude, in social and international organization; and stasigenesis, operating to limit progress and to keep old systems and attitudes, including even outworn superstitions, persisting alongside of or actually within more advanced social and intellectual systems. But there is an additional fourth sub-process, that of convergence (or at least anti-divergence), operating by diffusion —diffusion of ideas and techniques between individuals, communities, cultures and regions. This is tending to give unity to the world: but we must see to it that it does not also impose uniformity and destroy desirable variety.

As in the biological phase, major advance in the human phase is brought about by a succession of generally or locally dominant types. These, however, are not types of organisms, but of cultural and ideological organization. Monotheism as against polytheism, for instance; or in the political sphere, the beginning of one-world internationalism as against competitive multi-nationalism. Or again, science as against magic, democracy as against tyranny, planning as against *laisser-faire*, tolerance as against intolerance, freedom of opinion and expression as against authoritarian dogma and repression.

In broadest terms, the biological phase of evolution stems from the new invention of self-reproducing matter; the human phase from that of self-reproducing mind.

Man's acquisition of a second mechanism, over and above that of the chromosomes and genes, for securing both evolutionary continuity and evolutionary change, a mechanism based on his capacity for conceptual thought and symbolic language, enabled him to cross the barrier set by biological limitations and enter the virgin fields of psychosocial existence. By the same token he became the latest dominant type of life, shutting the door on the possibility of any other animal making the same advance and disputing his own unique position.

In the light of these facts and ideas, man's true destiny emerges in a startling new form. It is to be the chief agent for the future of evolution on this planet. Only in and through man can any further major advance be achieved—

though equally he may inflict damage or distortion on the process, including his own evolving self.

It is in large measure due to Darwin's work on biological evolution that we now possess this new vision of human destiny, and only by using Darwin's naturalistic approach in tackling the problems of psychosocial evolution can we hope to understand that destiny better and to fulfil it more adequately.

Evolution in the psychosocial phase is primarily cultural: it is predominantly manifested by changes in human cultures, not in human bodies or human gene-complexes. But, though it thus differs radically from evolution in the biological phase, the process is still a natural phenomenon, to be studied by the methods of science like other natural phenomena. Machines, works of art, educational systems, agricultural methods, religions, yes, and even men's values and ideals, are natural phenomena, at once products of and efficient agencies in the process of cultural evolution. The rise and fall of empires and cultures is a natural phenomenon, just as much as the succession of dominant groups in biological evolution.

The selective mechanism which determines what elements shall be incorporated and what rejected in the system of traditions, and so decides between alternative courses of cultural evolution, is primarily psychological or mental, involving human awareness instead of human genes, and directed towards the satisfaction of felt or imagined needs, instead of merely tending towards the survival of the more biologically fit: further, it operates only within the framework of human societies. We may call it *psychosocial selection*.

Though natural selection is an ordering principle, it operates blindly; it pushes life onwards from behind, and brings about improvement automatically, without conscious purpose or any awareness of an aim. Psychosocial selection too acts as an ordering principle. But it pulls man onwards from in front. For it always involves some awareness of an aim, some element of true purpose. Throughout biological evolution the selective mechanism remained essentially unchanged. But in psychosocial evolution the selective mechanism itself evolves as well as its products. It is a goal-selecting mechanism, and the goals that it selects will change with the picture of the world and of human nature provided by man's increasing knowledge. Thus as human comprehension, knowledge and understanding increase, the aims of evolving man can become more clearly defined, his purpose more conscious and more embracing.

Darwin ended *The Descent of Man* with this characteristic passage: "Man may be excused for feeling some pride at having risen, though not through his own exertions, to the very summit of the organic scale; and the fact of his having thus risen, instead of having been aboriginally placed there, may give him hope for a still higher destiny in the distant future. But we are not here concerned with hopes or fears, only with the truth as far as our reason permits us to discover it."

Today, building on the foundations provided by Darwinism, we can utilize evolutionary concepts in thinking about the history and future of our species. Human destiny need no longer be merely an affair of hopes and fears. In principle, it can be rationally defined on the basis of scientific knowledge, and rationally pursued by the aid of scientific methods. Once greater fulfilment is recognized as man's ultimate or dominant aim, we shall need a science of human possibilities to help guide the long course of psychosocial evolution that lies ahead.

HIGHER AND LOWER

IT is obvious to common sense that some organisms are *higher* than others—that a dog is higher than his fleas, or a fish higher than a jellyfish—but it is hard to define just what we mean by this. On the other hand, a proper definition is clearly important for general biological theory, because what appear to common sense as the obviously higher types of organism came later in the evolutionary process than the obviously lower types: mammals came later than reptiles; reptiles later than fish; vertebrates later than invertebrates; bees and ants later than cockroaches; flowering later than flowerless plants. The whole question is thus linked with the problem of direction in evolution in general, and with the idea of evolutionary advance or progress in particular.

A little reflection shows that "higher" and "lower" refer to levels or grades of organization. But then the further difficulty arises that organization itself is not easy to define scientifically. Herbert Spencer tried his magisterial hand at it, in the following remarkable sentence: "The law of evolution is as follows: an integration of matter and concomitant dissipation of motion, during which the matter passes from an indefinite incoherent homogeneity to a definite coherent heterogeneity and during which the retained motion undergoes a parallel transformation, the process continuing until it ceases in complete equilibration." This, like all of Spencer's attempts at definition, contains a great deal of truth; but under his attempt to generalize to the utmost, the concrete reality almost evaporates.

If one looks at the vast array of living organisms, I would say that *organization* denotes a structural and functional pattern in which differentiated parts are integrated into some unitary system of operation; any anatomist, physiologist, or biologist will be able to draw on innumerable examples from his own experience. Higher organization will then mean

35

more efficient integration of a greater number of more differentiated, more specialized kinds of parts.

What about the actual evolution of organization from lower to higher? It is true that higher organization, as I have just defined it, does arise during biological evolution, but it is also true that not all types of organism become more highly organized. For one thing, any new successful biological type will undergo what Rensch has called *cladogenesis* or branching evolution; it will branch or radiate out sideways in all sorts of specialized directions. For another, all types are eventually subject to what I have called *stasigenesis* or stabilized evolution. One type after another reaches a limit, after which it appears to be impossible for it to continue further in the direction of higher organization. For instance, echinoderms were never able to become truly bilateral; insects were severely limited in size by the nature of their respiratory system, and accordingly could never develop really large and plastic brains.

The way in which major improvement of organization actually occurs is through what palaeontologists term the succession of dominant types. During evolution some new type or pattern of organization appears which eventually proves its superiority to existing types, including that from which it itself has sprung, by multiplying at their expense, meanwhile undergoing cladogenesis and producing an adaptive radiation into many differentiated subtypes and specialized lines. Its evolutionary success proves that its new type of organization is higher: it is higher because it is more advantageous biologically. This is not arguing in a vicious circle; it is one of these virtuous circles which are so frequent in evolutionary theory.

As an example of the rise of a higher type of organization let me take the mammals as against the reptiles. There are today many more families and genera of mammals than existed in the late Cretaceous, but many fewer of reptiles. Reptilian organization is less efficient, notably in respect of temperature-control, of reproduction and care of young, and of brain structure and function. Accordingly, at the end of the mesozoic era the reptilian type became markedly reduced, many of its major subtypes becoming extinct. However, it survived, not in such abundance as the mammals to which it gave rise, but in considerable numbers of forms in those habitats to which it was well adapted. Indeed, as Darwin himself pointed out in the *Origin of Species* more than a century ago, we would expect some lower types to survive alongside the higher.

However, as he also pointed out in the *Origin* with remarkable acumen, natural selection will inevitably produce *improvement* (that was the term he used). It will lead to the improvement of most organisms in relation to their conditions of life; that is to say, we must expect natural selection to generate some sort of evolutionary advance.

This clearly applies to the evolution of specialized lines during the adaptive radiation of new major taxonomic groups such as Orders, as well as to the adaptation of single species or populations. It also applies to ecological communities. This latter point has been somewhat overlooked in evolutionary discussion. Any community of plants and animals must be adjusted to its environment and must be able to make efficient use of the resources of its habitat. In the course of time, under the influence of natural selection, it will become more efficient and better organized, until it eventually produces the maximum energy-flow possible in the conditions under which it operates. This was pointed out by Lotka years ago in his *Elements of Physical Biology*. There is what may be called an ecological metabolism of the whole community, just as there is a physiological metabolism of the individual species.

Thus when we consider evolving life as a whole, its differentiation, including the major differentiation into green plants, fungi, bacteria and animals, represents an extension of the phenomenon of adaptive radiation of a single type into many niches, combined with an extension of the phenomenon of more efficient exploitation of environmental resources by an ecological community.

This differentiation of life as a whole means a greater range of exploitation of more habitats or niches, from the deep sea, which was not occupied by vertebrates and higher molluscs until comparatively late, to hot springs, glaciers and the interior of other organisms; it also means a greater range of raw materials exploitable by life: there are nitrogen-fixing bacteria which manage to utilize atmospheric nitrogen, there are clothes-moths which manage to live on the keratin of wool or horn, there are termites which manage to live on wood. And do not let us forget that a higher level of organization may be achieved not merely by improving the organization of a given type, but by the combined organization of two complementary types, of which the most celebrated successful case is the symbiosis between fungi and algae to form lichens.

Metabolic exploitation may also be organized in time. The obvious case is that of metamorphosing animals, like frogs and higher insects, in which the creature exploits part of its environment for one portion of its life and a quite different part for another portion of its life.

Finally, let us remember that the improvement of metabolic organization during evolution will increase the variety and abundance of material resources for life to utilize, and these may then play a role in further evolution. Thus, in the early days of life's evolution there was no wood; when abundant wood was produced by plants, it provided the basis for a new type of exploitation by termites. Similarly, once terrestrial animals started producing the horny material we call keratin in bulk, clothes-moths and their relatives became an evolutionary possibility. Finally, when we come to man, we find that many substances produced by earlier life constitute important raw materials for his exploitation—coal, oil, and limestone, to name only three.

Improvement of organization may also be achieved through polymorphism, by the differentiation of types in the same species. The most obvious case is the differentiation between male and female in reproduction, which may of course be extended and elaborated; but polymorphism may operate by means of differentiated castes doing different jobs, notably in insect societies like those of bees, ants and termites, and resulting in a more highly organized social group.

38

The polymorphic differentiation of bee and ant societies clearly involves mind—mental activities of some sort. This reminds us that when we look at biological evolution as a whole, we find that the most notable improvement is the improved organization of mind; in other terms, a higher organization of the capacity for awareness. To take the example that concerns ourselves: man has become the latest dominant type in evolution because of his highly organized mind, the remarkable development of his mentally-accompanied behaviour.

In making this assertion we are brought up against the age-old and fundamental problem of the relation of mind and body, of the mental and the material. Let us begin with human beings, and recall that the only primary reality is the reality of our subjective experiences. We can only infer that other human beings have subjective experiences like our own. This is not only scientifically legitimate and necessary; it is justified pragmatically and operationally: human existence would be impossible unless we did so. In many cases, we can objectively check the accuracy of our inference that other human beings have subjective experiences similar to our own. We are sometimes able to obtain objective proof of deviations from that normal subjective pattern, deviations such as various types of colour-blindness, taste-blindness and smell-blindness. The subjective experience of such people must be slightly different from our own, though fundamentally similar.

Human beings are organizations of—do not let us use the philosophically tendentious word "matter", but rather the neutral and philosophically non-committal term translated from the German *Weltstoff*—the universal "world stuff". But our organization has two aspects—a material aspect when looked at objectively from the outside, and a mental aspect when experienced subjectively from the inside. We are simultaneously and indissolubly both matter and mind.

It is also both scientifically legitimate and operationally necessary to ascribe mind, in the sense of subjective awareness, to higher animals. This is obvious as regards the anthropoid apes: they not only possess very similar bodies

39

and sense-organs to ours, but also manifest similar behaviour, with a quite similar range of emotional expression, as anybody can see in the zoo; a range of curiosity, anger, alertness, affection, jealousy, fear, pain and pleasure. It is equally legitimate and necessary for other mammals, although the similarities are not so close. We cannot begin to understand or interpet the behaviour of elephants or dogs or cats or porpoises unless we do so to some extent in mental terms. This is not anthropomorphism: it is merely an extension of the principles of comparative study that have been so fruitful in comparative anatomy, comparative physiology, comparative cytology and other biological fields. It is equally fruitful when you extend it to the study of animal behaviour, this rapidly developing modern branch of science which is called ethology.

It is equally legitimate and necessary to extend the two-aspect concept to birds, which obviously have a high degree of both cognitive and emotional awareness, though it differs in many details from that of higher mammals; and it is quite legitimate to extend it, though clearly with decreasing sharpness and accuracy, to lower vertebrates and to some of the higher invertebrates. Here it is obviously more problematical, but I find it very difficult to deny some degree of aware experience to creatures like bees and ants. This poses a remarkable neurophysiological question: how can they have anything really comparable to our type of subjective and effective awareness in view of the immense difference in the size of their brains and the number of their neurones? That is a problem; whether we shall ever solve it I do not know, but it is a very important one. In spite of this vast qualitative difference, bees are capable, as we know from the remarkable work of von Frisch, of symbolic communication of information—one of the very few examples of non-human organisms with a true language as opposed to a set of signals.

Both the necessity of extending the capacity for aware experience to simpler forms of organization, and also the difficulty of drawing the line against still further extension, become even more obvious when we consider our own individual development. After all, we all start as a fertilized ovum, a single cell, and the development of a conscious creature from this must be gradual (unless you believe with some religious bodies that some special non-material entity called a soul is somehow inserted into the body at a definite period). Accordingly, there must be at least a potentiality of mind in the fertilized ovum; and when we return to phyletic development or evolution, the same must hold for an amoeba. In both ovum and amoeba we must postulate some mind-like quality, a mentoid (to coin an ugly but useful word) as well as a material aspect of their nature, some dim beginnings of subjectivity. However, this cannot become effectively operative or functional in their lives until they become equipped with a highly organized system subserving this very function, the ovum during the few months of its ontogeny, the primal amoeba during its billion-year evolution.

Biologists still have got to work out just how the non-functional, non-meaningful, "mentoid" activities of un-differentiated cells and organisms can be and have been summated and amplified by those special kinds of organization that we call brains, so as to become operationally effective in evolution. But it is beginning to look as if it depended on afferent impulses from a large number of different kinds of sense-organ, both extero- and intero-ceptors, being brought together and made to interact and become amplified and reinforce each other within single, organized, more or less closed, systems, without being able to issue directly in motor activity, as with reflex systems in the spinal cord.

But even should we discover the mechanisms by which mind becomes operative in evolution, it nevertheless remains true not only that the existence of mind is a fundamental fact but that the fact of its existence is the basic mystery.

What is the function of mind? Why did it evolve to increasing heights of intensity and importance? What is the biological value of the mental aspect of life in higher animals? It is now certain that natural selection through the differential reproduction of genetical variants is the essential agency of directional change in evolution. This being so, mind cannot be a useless epiphenomenon. It would not have evolved unless it had been of biological advantage in the struggle for survival. I would say that the mind-intensifying organization of animals' brains, based on the information received from the sense-organs and operating through the machinery of interconnected neurones, is of advantage for the simple reason that it gives a fuller awareness of both outer and inner situations; it therefore provides a better guidance for behaviour in the chaos and complexity of the situations with which animal organisms can be confronted. It endows the organism with better operational efficiency.

Primarily it generates quality out of quantity, and so helps animals, and ourselves, to discriminate more readily between different objects, events, and disturbances in the environment.

But equally important, the central machinery in the brain, the region where all the different sets of impulses from all the various sensory sources are brought together, provides for a fuller integration of our experiences.

This capacity of becoming aware of situations as wholes, of integrating a number of elements into a single experience, facilitates what the psychologists call insight behaviour, in which problems are solved, sooner or later, as wholes, not merely by repeated trial-and-error learning or conditioning.

Just as the evolution of new kinds of material metabolism can provide the basis for further biological change, so the evolution of new kinds of awareness has consequential effects on further evolution. For example, as soon as organisms had come into existence endowed with the capacity for coloured pattern-vision, new colour-patterns began to evolve, sometimes in all members of the species, or in members of the opposite sex, or in other species in the same ecological community. Such new colour-patterns, whether they subserve warning or display or cryptic functions, are called allaesthetic, because they have developed in relation to the visual capacities of the other individuals or species. Thus the whole evolution of striking colour-patterns in sexual display in birds is consequential on the evolution of colour-vision in the class Aves. The absence of pure scarlet in the flowers of bee-pollinated plants is a consequence of the bees' total red-blindness, while their capacity for seeing ultra-violet has led to the evolution of floral patterns which are invisible to our eyes.

The evolutionary development of such consequential colours and patterns is of extreme interest. It takes place primarily through the evolution, under the influence of natural selection, of rather crude patterns of stimulation called sign-stimuli (or stimulus-patterns), which release specific patterns of behaviour in other individuals. The work of men like Lorenz, Tinbergen, Thorpe, Hinde and J. Z. Young has shown that in vertebrates, in higher insects and in cephalopods there are a number of these comparatively simple but highly distinctive stimulus-patterns which interact with neural mechanisms in the brains of other organisms, mechanisms which release a certain motor pattern of action, as a key works in a lock. For this reason these sign-stimuli are often called releasers.

For instance, newly hatched pesserine birds will automatically gape for food when confronted with any smallish round object; and Heinroth records that many birds in zoos will react to the first appearance of Swifts in spring as if they were Hawks. However, individual learning may be superposed on such built-in mechanisms and override their automatism. The zoo captives soon learn that Swifts, in spite of their shape and speed, are harmless, and will not crouch or try to escape at sight of them; and the songbird nestlings gradually learn discrimination and eventually come to distinguish their individual parents from other adults.

Similar releasers have been operative in the evolution of courtship and threat displays in birds, as I have set forth in another essay in this volume.

In some birds, display characters and activities are exaggerated and their stimulative value presumably enhanced. This occurs when the selective advantage of reproductive success is very high, as in polygamous-promiscuous species, in which one male may make ten or twenty successful matings, as against two, one or even none by rival males. In such cases, display characters are enhanced to an extraordinary extent, until they interfere with the ordinary

43

activities of life, as you may see in the male Peacock, or still more in the male Argus Pheasant.

The apparent high stimulative value of such exaggerated display characters leads on to another extremely interesting fact—the existence of subnormal and supernormal stimuli. Many species of birds, if their eggs are removed from the nest, will roll in eggs which are placed near by, whether they are real eggs or painted plaster models. Indeed, many objects within a certain range of size and shape will stimulate rolling-in activity, though such abnormal objects generally act as subnormal stimuli. However, a model egg double the normal size acts as a supernormal stimulus, and will be eagerly and repeatedly rolled in if replaced outside, even though when the bird has got it into the nest it cannot manage to brood it properly, but keeps on slipping off it sideways.

These facts have various applications in the evolution of our own species. Although very few simple sign-stimulus patterns operate as releasers in man, there is one of great importance—the smile. Very young infants will respond to their mother's smile; they will also respond to a crude drawing of a human face, provided that the sides of the "mouth" turn up. The smiling face pattern acts as a releaser to a built-in response. In point of fact, the baby's response is also a smile, and this tends to provoke a similar response in the mother. Thus the mechanism is self-reinforcing, and has an important function in establishing an emotional bond between mother and child. In later life, of course, the built-in mechanism is supplemented and improved, made more subtle and complex by learning.

Then let us take colour-pattern and its releaser function. In sub-primate mammals we do not find the bright colours that occur in birds; their range of coloration is restricted to blacks, whites, browns, yellows and greys; you do not find any pure blues, pure reds or pure greens. This presumably is because all sub-primate mammals appear to be more or less colour-blind. Primates, on the other hand, including ourselves, are capable of colour-vision: accordingly in them the evolution of bright colours has occurred. It has sometimes occurred in peculiar places: we need only think of the

Mandrill, in which not only the face but also the bare rump is brightly coloured and conspicuously displayed in certain situations. P. G. Wodehouse referred to the Mandrill as "an animal that wears its club colours in the wrong place", and R. L. Stevenson wrote of the "blue-behinded ape". It is a noteworthy evolutionary fact that colour comes into its own again as a sign-stimulus as soon as the stock re-acquires the capacity for colour-vision.

In ourselves, too, colour-pattern and visual form play an important role as in lips and cheeks and the shape of women's breasts. But man, in addition to or in place of built-in releasers, has the possibility of deliberately creating symbols which then can function as artificial sign-stimuli, and can act as learned releasers for various actions or emotions. The most obvious case is that of national flags, which stimulate patriotism. The swastika and the cross are examples of symbols functioning as ideological and religious sign-stimuli. But the most important effects of this human capacity to create significant symbols are seen in the sphere of communication: in this field man has created various signs and symbolic constellations of stimuli, including the whole system of true or symbol-using language, which can be used to communicate a virtually unlimited range of external and internal situations.

The development of higher degrees of awareness in animals leads on to the beginnings of what we must call *tradition*, the handing on of the results of individual experience directly from one generation to the next in addition to the indirect and cumbersome genetic method operating by natural selection and based on the "fossilized" genetic information stored in the chromosomes. Tradition across the gap of a single generation is found quite often in the carnivora, like foxes or lions: the young learn to hunt better by accompanying and learning from the adults. In one or two other cases we find the beginning of a continuous incipiently cumulative tradition. The best example is that of a Japanese monkey. In this species, each troop has its own food-traditions, which are slightly different from those of every other troop. These different traditions appear to be learned,

45

or largely learned. Most interesting of all, now and again
there is a change in a troop's food-tradition; this apparently
always stems from a single young monkey who in our human
terminology would be called naughty, who ventures to eat
some forbidden fruit or tabooed vegetable. His mother
punishes him, but he goes on eating it because he is naughty
and because he likes it. Eventually the mother takes a taste
too, and then the new habit may spread and eventually
becomes incorporated into the tradition of the troop.

In other cases, evolution can bring about an increase in
potentialities, which, however, are not actually realized in
the normal life of the species, but may be realized in abnormal
or new circumstances. To put the matter in another way, the
acquisition of a certain pattern of organization may have
consequential results, results which it did not originally produce, but which may be produced in certain new conditions.
I will take two examples of such consequential results in
birds, arising from their general capacity for discrimination
and learning. First, Professor Otto Koehler some time ago
proved that certain birds, such as Jackdaws, have a capacity
for non-symbolized (non-numerical) counting which is just
as well developed as yours or mine, though they never make
use of it in their normal lives. They can distinguish between
groups of five, six and seven objects, even if the objects are
heterogeneous and arranged in different patterns; this is just
about as much as the average human being can do without
having recourse to verbal or mental counting. My second
example is a remarkable new food habit acquired by the
common Blue Tit in Western Europe, including Great
Britain, in recent years. Over a large area, Blue Tits have
acquired the habit of opening milk bottles put outside the
door and taking the cream. This has been shown by an
elaborate survey to be the result of a series of a few individual
"inventions": the habit started in a small number of different
localities and then spread from these foci. There must be a
few tit geniuses who first discovered how to open the lids of
the bottles, after which the habit spread, either by direct
imitation or more probably by semi-imitation.

In our own species we have one of the most striking

46

examples of a consequential effect in the capacity of the human mind (perhaps not all human minds, but a considerable number) for understanding and practising higher mathematics. Certainly our brains were not evolved in order to be able to solve differential equations—that would have been of no use to our primitive proto-human ancestors. Mathematical capacity was a consequential potentiality of the way in which human brains are able to organize awareness in general, and the potentiality is actualized when civilization and culture reach a certain level.

The origins of new capacities are always interesting. For instance, the beginnings of art are found in Bowerbirds, which adorn the entrances to their bowers with bright objects. Some species have definite colour preferences; thus the Satin Bowerbird prefers blue objects, but removes red ones if placed by the bower. The Satin Bowerbird also "paints" the lower part of its bower with the juice of berries mixed with charcoal. But here individuality plays a role: only certain males do this, while others leave their bowers unpainted.

One of the most interesting phenomena of biology is the appearance during evolution of activities which are enjoyed for their own sake. Anybody who has ever watched birds closely will know that a number of species indulge in aerobatic "flight games", which they enjoy just as the human species enjoys skiing or skating. Adelie Penguins, as Levick sets forth in his book *Antarctic Penguins*, indulge in "joyrides" on ice-floes drifting past their rookery in the current, and will swim back to repeat the ride on another floe. Mammals too enjoy various kinds of physical sport; I think of seals, porpoises and, above all, otters. Otters even invent sports: after a snowfall they will run over the crest of a little rise, then turn on their backs and enjoy an inverted slide down the slope.

There are other animal potentialities which are not normally realized. Last year I was fortunate enough in Africa to see the famous lioness, Elsa, leap into the clearing where we were camped, to greet Mrs Adamson: it was just like two devoted human friends meeting after an absence. Mrs

47

Adamson had elicited out of this wild creature something in
the nature of an integrated and affectionate personality.

Many other latent potentialities of higher mammals have
eventually been elicited into actuality. Some of the most
important consequential effects are the result of an animal's
ability to manipulate objects, which is always coupled with
what ethologists call the exploratory instinct, the urge to
explore your environment. This is especially strongly
manifested in the higher primates, the chimpanzee and other
anthropoid apes. As a consequence, they can learn to perform
the most extraordinary physical feats, on bicycles, on pony-
back, on roller skates. A further consequence is their
capacity to integrate a number of experiences in one act of
awareness. This has led to the development of what the
psychologists call *insight*, the solving of problems not merely
by trial-and-error but all at once, like fitting two short sticks
together, one hollow and one solid, to make a longer one.

The same sort of development of skill and intelligence
occurs in all organisms which are capable of manipulating
objects.[1] This applies to elephants, though as they use their
trunks to handle objects I suppose that *trombipulate*, not
manipulate, is the correct word. (I am sure that Dr Johnson
would have agreed.) The trombipulative capacity of the
elephant is certainly coupled with a striking development of
awareness and an astonishing subtlety of social behaviour.
For instance, in the Murchison Falls National Park in
Uganda my wife and I saw the skull of a famous elephant
called Methusaleh because of his great age. For some years
before his death he had lived separately from the herd. But
he did not live alone, for he was constantly accompanied by a
young bull who acted as a kind of squire, keeping him company and protecting him against intrusion. This, we were
told, is the normal course of events when an old herd bull
starts living a separate life.

Another elephantine incident, though irrelevant to my
present argument, is I think worth mentioning here. When
my wife and I were visiting this same Murchison Falls Park,

[1] The skill and intelligence of porpoises and dolphins appears to be based
on their capacity to "manipulate" their entire body.

the Warden told us that some months earlier he had heard
an unusual type of excited screaming by an elephant. He
came around a corner to find a middle-aged elephant bull
approaching a younger male with homosexual intentions.
The younger male was rejecting these improper advances;
eventually the older bull became so frustrated that he lay on
his back, rolled on the ground, and trumpeted. The younger
male thereupon sat down on his haunches and just looked at
the other; upon which the Warden laughed so loud that both
the elephants took fright and ran off.

To sum up, during biological evolution the level of
organization of awareness was raised. This led to a steady
increase, extension and elaboration of what we may call the
animal's significant world, that part of the environment
which has relevance or meaning for it. Think of what the
world means to an amoeba, what it means to a worm and to a
fish, then to a dog and then to ourselves, you will see how
important this extension of the organism's significant world
has been.

I may put it in another way. Throughout evolution the
animal metabolizes with the aid of its body organs, it transforms the raw material of its food, drink and oxygen into
characteristic biochemical patterns which canalize and direct
its physiology. But with the aid of its brain, the organ of its
mind, it metabolizes the raw material of its subjective
experiences and transforms it into characteristic patterns of
awareness, which then canalize and direct its behaviour.
This we may call *psychometabolism*. During the late stages of
evolution increasingly efficient psychometabolism is superposed on universal physiological metabolism.

After some two and three quarter billion years biological
evolution, as operated by natural selection and as manifested
in the improvement of purely biological and physiological
functions, seems to have reached a limit. The only road to
further advance was through psychometabolic improvement,
via brain and mind. There was a major breakthrough to
something wholly new—the self-reproduction of mind and
its effects. This was achieved through the development of
true symbolic language, which permitted the cumulative

49

transmission of useful experience, and initiated a new phase, mode, and method of evolution, the psychosocial. As a result, evolution ceased to be primarily organic, dependent on genetic change, and became primarily and increasingly cultural, dependent on cultural change. Psychometabolism began to dominate life, and its products came increasingly to direct the new artificial forms into which man builds the raw material of his physical environment.

In this new phase of evolution there is no longer any distinction between soma and germ-plasm as there is in the biological phase: the whole psychosocial system evolves and is itself transmissible; it is both formative and formed. As a result, concepts and beliefs gradually evolve, and may become more relevant to the changing conditions of life. On the other hand, concepts and beliefs may persist and continue to affect man's life even when, in the light of later experience, they can be seen as false and erroneous. An obvious example of false beliefs affecting the course of events was the belief of the Aztecs that the sacrifice of human beings was necessary to ensure that the sun should continue to rise each day. Accordingly, more and more prisoners were captured to be sacrificed to the sun. Not unnaturally, the neighbouring tribes did not like their men being used in this way, and accordingly most of them sided with Cortez. Thus the Aztecs' erroneous beliefs contributed powerfully to their downfall.

The persistence of demonstrably incorrect ideas is illustrated in Communist Russia. Marx laid down that communist revolution would start in the most highly industrialized countries, would be implemented by the dictatorship of the proletariat, and would lead to a classless society and the withering away of the State. Although the actual course of events has been entirely different, yet these ideas are still extremely potent and continue to affect the actions of a great many people in the world.

Another important result of evolution's entry on its new, psychosocial phase is that consciousness, including conscious purpose, can now take a hand in the evolutionary process. The purpose that many philosophers and theologians like to think that they can discern in nature, and that Archdeacon

50

Paley believed that he had conclusively demonstrated, turns out to be merely an apparent purpose, the result of the wholly non-purposive workings of natural selection; only in psychosocial evolution can true purpose begin to operate. Natural selection is an ordering mechanism which orders the process of biological evolution in relation to the differential survival of biologically more efficient types, whereas psychosocial selection is an ordering mechanism which orders the process of psychosocial evolution in relation to conscious or unconscious aims directed towards greater enjoyment and fulfilment.

In man, there are dominant systems of ideas which guide thought and action during a given period of human history, just as there exist dominant types of organisms during a given period of biological evolution. After a certain time, a dominant organization of thought may no longer fit the developing conditions of human life, or may come up against a limit and find that its capacity for interpreting the world and providing comprehension of human destiny is inadequate. Then history has to wait until a new and more appropriate organization of ideas and beliefs is brought to birth and becomes the new dominant system.

We are now on the threshold of some such critical revolution of thought in which general human ideology is destined to be radically reorganized, and our old patterns of ideas and beliefs will be superseded by a new dominant idea-system.

Today, as Sir Charles Snow has pointed out so forcibly, there is an unfortunate split between our Two Cultures—the scientific and the humanistic culture—which has now been superposed on the good old split between science and religion. This dual split is largely maintained by the resistance of many humanists and philosophers and religious people to the full extension of science and the scientific method into the human or psychosocial sphere; and also, I must regretfully admit, by the inhibitions of many natural scientists and their disinclination to see science trying to tackle problems which lie outside the range of established scientific activity today. However, this extension of science, or rather the interpenetration of scientific and humanistic

51

method, has now become urgently necessary. If we look at past history we see that most of the spectacular changes in human life in the last four centuries—many of them definitely worthy to be called advances—have been due to the spread of the scientific approach and scientific method into new and increasingly complicated fields and subjects, with the enlargement and strengthening of man's basis of established, tested, and well-organized knowledge. Starting with terrestrial and celestial mechanics, science progressively invaded the fields of physics, chemistry, geology, anatomy, physiology and general biology, and recently, though still in a rather modest way, psychology and the social sciences.

The next major step must be for the scientific method to be brought into the core of the psychosocial sphere. Man must take a scientific look at his values, at his ethics, at his art and aesthetics, at his social and economic organization, and at his religion. Above all, it is necessary to take a scientific look at the historical process in general, as we have successfully begun to do with the general process of biological evolution. In so far as we succeed in this new scientific venture, we shall be able to construct the framework for a new, open-ended, and much more comprehensive pattern or system of ideas, capable of expansion and application in many new fields.

PSYCHOMETABOLISM

A COUPLE of years ago I agreed, not without some trepidation, to address a gathering of psychiatrists on some general psychological topic. I eventually decided to approach the subject in the general perspective of evolution, and to speak about the role of mind as an operative factor in the evolutionary process.

If we look at the process of biological evolution as a whole, we will see that it tends towards the production of types which can utilize more of the world's material resources more efficiently. To achieve this, the processes of physiological metabolism are improved, and new types of metabolic utilization appear.

The other major tendency in biological evolution is manifested in the evolution of mind, a trend towards a higher degree of awareness. This is especially marked in the later stages of the process in the dominant types of animals, notably insects, spiders and vertebrates, and is of course mediated by their brains. Brains can be regarded as psychometabolic organs. Just as the physiological metabolic systems of organisms utilize the raw material provided by the physico-chemical resources of the environment and metabolize them into special material substances, so brains utilize the raw materials of simple experience and transform them into special systems of organized awareness.

This at once brings up the problem of the mind-body relation. Here, as I have pointed out in the last chapter, the evolutionary approach is essential. Let us begin with man. The human organization has two aspects: first, a material one when seen from the outside, and secondly a mental or subjective one when experienced from the inside. It is obvious that we must ascribe subjective awareness to higher animals, as Darwin did in his great book *The Expression of Emotions in Man and Animals*. It is all too obvious for the higher apes. It is equally legitimate to say that mammals such

53

as dogs must possess a marked degree of subjective awareness; otherwise, indeed, we should not be able to interpret
their behaviour at all. We can extend the principle to lower
vertebrates with a high though lesser degree of certainty.
Indeed, I do not see how we can refuse some sort of subjective awareness to higher invertebrates.

The legitimacy of extending the capacity for awareness to
less complex organizations than our adult human selves is
equally obvious when we consider our own development. We
all start as a fertilized ovum whose behaviour gives no
evidence of awareness, so our capacity for subjective awareness must arise naturally and gradually in the course of
development out of some dim original potentiality.

In this connection an analogy with bio-electricity is useful.
As any zoologist knows, there are several genera of fish,
which possess electric organs with definite functional utility;
they help their possessors to find their way about the muddy
waters where vision is not of much use, or to give paralysing
shocks to their enemies or prey. In parenthesis, you will find
that Galen records the first employment of electro-therapy—
the use of electric shocks from torpedo fish to cure headaches.

These electrical properties of certain fish were for a long
time considered as something unique in biology, and the
problem of their evolution was a great puzzle to Darwin
himself. Eventually, however, the physiologists found that
every activity of the body—muscular contraction, nervous
conduction, sensory reception, glandular secretion—certainly
in all mammals and presumably in all vertebrates, is accompanied by electrical changes in the organs concerned. These
are without adaptive significance in the life of the animals,
and are merely by-products, consequences of the way protoplasm is made; in the same way, the red colour of blood had
no original function but was simply a consequence of the
chemical composition of vertebrate haemoglobin. In higher
primates, the red colour of blood has ceased to be a mere by-
product and has acquired a signalling function through the
evolution of thin-skinned lips and other organs of the body:
in a somewhat similar way in electric fish special electric
organs have been evolved, particular organizations of tissue

54

which summate and amplify these minute electrical accompaniments or epiphenomena of protoplasmic activity to the point where they can be, and are, functionally effective and biologically valuable.

This provides a perfectly good analogy with the evolution of mind. In this view, every living organism has what I may call a "mentoid" or potentially mental aspect, something of the nature of subjective awareness which is merely a consequence of the way it is made, and confers no biological advantage in its life. Brains, on the other hand, are mechanisms for intensifying, amplifying, and organizing life's original dim subjectivity to a point where it can properly be called *mind*, and becomes significant in the animal's life.

How does it do this? For one thing in some unexplained way, it generates qualitative distinction out of quantitative difference. The sensation of blue is irreducibly different from red. The difference between blue and red depends on quantitative differences in the frequency of the light-waves reaching the retina and of the impulses passing up the optic nerves, but as sensations, blueness and redness are qualitatively distinct. Biologically, this permits readier discrimination between objects: it is much easier to discriminate between two qualitatively different colours than between two quantitatively different shades of grey. This is, of course, the basis of tests for colour-blindness.

Discrimination is similarly aided by the radical qualitative differences between the different modalities of sensation— sight, hearing, touch, smell, and so on—which again are irreducible in terms of any common factor. Again, it is essential to be able to discriminate potentially damaging situations and objects from those which are potentially enjoyable and useful, and this has been achieved through the radical qualitative difference between the sensations of pleasure and of pain. Similarly, it is valuable to discriminate between threatening or dangerous situations and desirable or useful ones: this has been achieved by the evolution of sharp quantitative differences in our "built-in" emotions— fear as against curiosity, for example, or sexual attraction as against hostility.

55

Finally, the central organ of awareness, the brain, has the astonishing capacity of integrating an enormous number of separate, and often disparate, elements of experience into an organized pattern of which the animal is aware as a whole, and which it experiences as different from all other such patterns. Our perceptions are not immediately given; they are not merely, so to speak, photographed automatically on to some sensitive mental film; they are built up in the course of our early life through a combination of our visual and tactile and other experiences in a kind of automatic learning process. In addition, there is the integration of different experiences in time, by conditioning and by learning in the customary sense, resulting in various forms of memory. As a result of such processes, animals which stake out a definite territory can integrate all sensory information concerning the territory into a single field of experience.

This integration of sensory information into organized patterns which can be readily discriminated in awareness may produce extraordinary results. Some of the most extraordinary are concerned with the way in which animals find their way about. For instance, we now know that migrating birds find their way by steering with the sun if they are day migrants and by the constellations if they are night migrants. Of course, they can only do this by means of some extremely elaborate computer system in their brains—though this is no more elaborate than the computer system in our own brains which enables us, while playing tennis, to anticipate our opponent's stroke with appropriate movements of our own. In both cases, however, the computer system is only the mechanism of the action; the organism must in some way be aware of the situation as a whole in order to put the computer system into operation. This instantaneous awareness of total situations is a psychometabolic activity of decisive importance for successful behaviour.

Many interesting psychometabolic organizations operate in higher animals. Frequently learning capacity is grafted on to an innate response. A good example is found in the English Robin, which is quite different from the American Robin, a smaller bird with a brighter red breast. In the

56

breeding season, the sight of a red-breasted rival will stimulate hostility in a male Robin in occupancy of a territory. The same effect is produced by a stuffed dummy; even if the dummy's head, tail and wings are removed, leaving only a fragment of body with a patch of red feathers, this will be attacked in the same way. The red breast is a specific releaser of behaviour. The sight of it is a simple sign-stimulus which releases a built-in mechanism of attack. On the other hand, the system can be modified by the further psychometabolic activity of learning. A male Robin will learn to accept a female as mate even though she too has a red breast; and eventually, through becoming aware of slight differences in behaviour, he will learn to discriminate between his individual mate and other female Robins, though they are indistinguishable to the human observer.

These simple visual patterns serving as releasers of specific behaviour, have come into prominence through the work of men like Lorenz and Tinbergen and Thorpe. Thus the newly hatched Herring Gull pecks at its parent's showed bill, which then regurgitates food for it. Tinbergen that if the newly hatched young is tested before it has even seen an adult bird, it will peck just as well at a coloured cardboard model as at the real parent's beak. The pattern of the beak—yellow with a red spot at the tip of the lower mandible—acts as a sign-stimulus operating the mechanism releasing the pecking reaction.

The whole system is "innate"—genetically determined. But its operation can be modified. The normal beak is yellow, fairly elongated and with a red spot near the end of the lower mandible. A model without any spot has very little effect on the young. One with a spot of another colour than red will be less efficient than the normal pattern, but more so than a model without any spot. Shortening the beak will make the model less effective, while a model which is not in the least like a normal beak, but is a very elongated rectangle with a very striking pattern near its end, will elicit a *supernormal* response—it will induce the young to peck at it more vigorously than they will at their own parent's beak.

We find similar phenomena in human beings. Here again,

57

we find allaesthetic characters. In the evolution of man both colour and form have been employed as sign-stimuli, releasing sexual behaviour or at least promoting sexual attraction. The red colour of lips and cheeks is obviously of value in sexual selection, and its supernormal enhancement by rouge and lipstick is the basis for a large portion of the lucrative cosmetics industry. The form of the female breast is also a sexual sign-stimulus: its enhancement is the basis for the manufacture of brassieres, and its supernormal exaggeration has given rise to the article known as "falsies". My American friends tell me that falsies are now obtainable for the other side of the female anatomy—false bottoms, in fact. If you want really striking examples of supernormal stimuli, you have only to think of Marilyn Monroe or Sophia Loren.

Ethology, as the study of animal behaviour is now called, has led to some illuminating facts about the evolution of display in the courtship and rivalry of birds, and the role of releasers in the process. Display appears to originate in what the ethologists call intention movements. For example, a bird may see an approaching individual as a potential enemy or rival. Its first action is to make ready for attacking if need be, by assuming an attitude preparatory for fighting, an attitude of hostile intention. On the other hand, if it is frightened its first reaction will be to prepare for withdrawal, an intention movement directed towards flight. In the course of evolution such intention movements, whether simple or the result of two conflicting drives, may be seized on by natural selection and ritualized, fixed in the hereditary constitution and converted into releasers in their own right; and their effectiveness as releasers may then be further enhanced by the development of special plumage and brilliant colour. This is perhaps most clearly seen in the evolution of threat displays. The effectiveness of the hostile intention movement of an angry Lion has been enhanced by the evolution of a mane; the effectiveness of the Drill's intention action in baring its teeth has been enhanced by the evolution of ivory-white gums and of white patches above the eyes which are revealed by muscular action.

Extremely interesting things happen when there is a

58

conflict between two contradictory drives or emotions. For instance, birds are apparently always afraid of too close an approach by or to any other individual. So when, at the beginning of the breeding season, a male is attracted by a female and starts to approach her, as he gets really close to her he will begin to be frightened and to show hostility. There is thus a conflict of emotional drives, and this results in a compromise between two types of intention movement, one of approach and the other of retreat. These compromise attitudes may again in the course of evolution be "ritualized", as the jargon is, and converted into releasers in their own right, and then polished up by natural selection, with their effectiveness enhanced by further stimulative adornment. Many display actions in birds have originated from such compromise attitudes.

One of the most interesting discoveries of modern ethology is the fact that when two conflicting drives operate simultaneously at high intensity (and in certain other cases) some apparently quite irrelevant activity is produced. The two drives cancel each other out, so that no specific actions appropriate to either of them are performed. However, there has been a general raising of the excitation-potential in the brain, and this spills over into whatever other channel is available, resulting in what has been called a *displacement activity*. In birds, a conflict of this sort frequently spills over into displacement preening, though the act of preening is quite irrelevant in the context of the situation. Furthermore, in various groups of birds, notably in the ducks, the action of displacement preening itself has been ritualized; it has been made more conspicuous by the evolution of appropriate colour and form, and converted into part of the organized display-activities which serve to bring about successful mating. This reaches its culmination in the Mandarin Duck. In Tinbergen's book on instinct you will find other examples of this remarkable phenomenon, as well as a detailed presentation and discussion of the nature and function of releaser sign-stimuli in general.

As Tinbergen and Lorenz have shown, these facts have relevance for man. Human beings show many displacement activities, such as scratching their heads when puzzled. More basically, conflict and the reconciliation of conflict in meaningful activity are of fundamental importance in human mental development, and are one of the chief concerns of psychiatrists. In man, instead of conflicting drives resulting in overt compromise attitudes, they often continue to operate internally. This results in a conflict of what we may call "intention urges"—urges or drives towards aggression, or fear, or sexual attraction; but whatever the conflict is based on, it is not overtly manifested in action. The problem is this: can these conflicting urges be reconciled internally and converted into something which combines the energy of both drives in a single and functionally valuable "super-drive"? Or are their energies going to remain locked up, so to speak, in functionally useless conflict? Or is half the energy of the conflicting drives going to be wasted by the repression of one of them?

In man there are very few examples of built-in sign-stimuli acting as releasers. The best-known is the so-called smile reaction. As I have already mentioned, this is a self-reinforcing process. When the infant smiles at the mother, even if she is not smiling, she will smile back in return, and vice-versa. The self-reinforcing process establishes and helps to strengthen the emotional bond between mother and infant.

This establishment of emotional bonds between members of a species is obviously of the greatest importance in evolution. Once again there are traces of it in sub-human creatures. My first important piece of behaviour study was on the courtship behaviour of a British bird, the Great Crested Grebe. In this species both sexes develop elaborate sexual adornments in the breeding season and employ them in mutual displays. It is quite clear that, in addition to their stimulative function, these displays, however they originated, now serve as an emotional bond between the members of the pair. They serve to keep the pair associated throughout the whole season, during which the young need to be looked after by both parents.

The bases of such emotional bonds are sometimes very

interesting. In his fascinating studies of monkeys, Professor Harlow of Wisconsin took new-born monkeys away from their real mothers and gave them pairs of surrogate mothers. Both possessed an iron framework and a crude model of a face; one of them, the feeding mother, was provided with a bottle of milk which was the baby monkey's only source of food; the other, or furry mother, merely was covered with a furry material. In contradiction to psychoanalytic theory, the baby monkey chose to spend much more time with the furry mother, who did not satisfy its hunger but gave it a feeling of protection and an agreeable tactile sensation. When it was kept shut up in a room with windows, it would console itself by gazing at the furry mother, not the milky mother. It will be extremely interesting to see what happens to these monkeys after they have been brought up entirely by arti-ficial mother surrogates. Can they be made to reaffirm an emotional bond with a real female monkey or a human surrogate?[1]

Besides bonds between members of a mated pair, and those between parent and offspring, there are familial bonds and the extremely interesting social bonds that operate in organized animal societies. Konrad Lorenz's delightful and important book, *King Solomon's Ring*, gives an account of some of these. I have only time to mention one, but one which is of great interest. The wolf-pack is an organized society of proverbially aggressive animals, but when a bigger or higher-ranking wolf is quarrelling with a smaller or younger one, and the smaller one feels that he is in danger of being beaten and hurt, he will adopt a special "appease-ment attitude", deliberately displaying his most vulnerable part. This acts as a sign-stimulus which definitely inhibits further aggressive behaviour on the part of the larger wolf, or, if you prefer, releases a non-aggressive pattern of behav-iour. However angry he may have been, he just finds himself unable to go on attacking the smaller animal which is

---

[1] It has later been established that such monkeys suffer irreversible emotional damage and grow up permanently abnormal. Something of the sort occurs with human babies deprived of adequate maternal love and care during a critical period of their infancy, as John Bowlby has shown.

advertising its defencelessness. An ethologist friend of mine has applied this fact to human situations. He has twice recently avoided punishment for motoring offences by assuming a cringing self-deprecatory "appeasement attitude". In one case an aggrieved car-owner didn't even take his name and address; in the other a policeman wouldn't give him a ticket. I recommend this as a very useful piece of applied psychology, but it needs histrionic skill.

Finally there are the bonds between generations. These become of increasing importance in higher vertebrates. In some birds and mammals we see the beginnings of what one must call tradition, the handing down of the results of experience from one generation to the next. Originally this occurs only across a gap of one generation; but in Japanese monkeys each troop has its own food-tradition. This is the real beginning of culture in the anthropological sense, based on the cumulative transmission of experience, including some novel experiments.

Now at last I can tackle my real subject. Throughout evolution, the animal, with the aid of various bodily organs, utilizes the raw materials of its food, drink, and inspired air and transforms them into characteristic biochemical patterns which canalize and direct its physiological activities. This is metabolism. But with the aid of its brain, its organ of awareness or mind, it utilizes the raw material of its subjective experience and transforms it into characteristic patterns of awareness which then canalize and help to direct its behaviour. This I venture to call psychometabolism.

During the latter stages of evolution, an increasingly efficient type of psychometabolism is superposed on and added to the universal physiological metabolism. Eventually, about ten million years ago, purely biological evolution reached a limit, and the breakthrough to new advance was only brought about by the further elaboration of the psychometabolic apparatus of mind and brain. This gave rise to man: it endowed him with a second method of heredity based on the transmission of experience, and launched him on a new phase of evolution operating by cumulative tradition based on ideas and knowledge. Both novelties are of course superposed on the biological methods of transmission and evolution, which he also possesses.

In man, organizations of awareness become part of the evolutionary process by being incorporated in cultural tradition. Accordingly, in human evolution totally new kinds of organization are produced: organizations such as works of art, moral codes, scientific ideas, legal systems, and religions. We men are better able to evaluate, to comprehend, to grasp far more complex total patterns and situations than any other organism. We are capable of many things that no other animal is capable of: conscious reflection, the idea of self, of death, of the future in general; we have the capacity of framing conscious purposes which can then be translated into action, and of constructing values as norms for our activities. The result is that evolution in the psychosocial phase is primarily cultural and only to a minor extent genetic.

Of course all these new types of organization evolve like everything else. The science of comparative religion shows how religions have evolved and are still evolving. The history of science studies the evolution of scientific ideas and how they become operative in the psychosocial process.

Our mental or psychometabolic organizations fall into two main categories: those for dealing with the outer world and establishing a relation with external objects; and those for dealing with our inner world and relating our perceptions and concepts and emotional drives to each other and integrating them into a more or less harmonious whole. The ultimate aim is to deal with all kinds of conflict and to reduce mental friction, so as to get the maximum flow of what is often called mental energy.

Here I want to put in a plea against the physicists' bad semantic habit of appropriating terms from common human usage and restricting their employment to physico-chemical phenomena. In the strict physicist view, it is no longer permissible for a biologist to use the term "mental energy"; for the physicists, energy is something exclusively physical and mathematically definable in terms of mass, velocity and the like. But the biologist and the psychologist also need a terminology. There *is* something operating in the awareness-

63

organization of man and higher animals which is analagous to energy in the physical sense, and can operate with different degrees of intensity. For this, we may perhaps use the term *psychergy*, without committing ourselves to any views as to its precise nature.

A major job for all disciplines concerned with human affairs, whether biochemistry, psychology, psychiatry or social anthropology, is to investigate the extraordinary mechanisms underlying the organization and operation of awareness, so as to lay the foundation for and promote the realization of more meaningful and more effective possibilities in the psychosocial process of human evolution.

When we look at animal behaviour it is clear that differences in possibilities of awareness between different species are primarily genetic. One species of bird prefers blue, another does not: the sign-stimulus which will release adaptive patterns of action in one species of bird will not do so in another. There is obviously a genetic basis for the difference.

Equally obviously there is a genetic basis for the difference between the genetically exceptional individual and the bulk of the species. All great advances in human history are due to the thought or action of a few exceptional individuals, though they take effect through the mass of people and in relation to the general social background. We have seen how, already in birds and mammals, the exceptional individual can be of some importance in the life of the species. In man, the exceptional individual can be of decisive importance.

Today, many workers in psychology and psychiatry and other behavioural and social sciences resist or even deny the idea that genetic factors are important for behaviour. They are undoubtedly wrong. Of course environmental factors, including learning, are always operative, but so are genetic factors. To take an example, genetic differences in psychosomatic organization and somatotype are obviously correlated with differences in temperament, and these with different reactions to stress and proneness to different diseases.

Frequently, it is not so much complete genetic determination we have got to think about, but rather proneness to this or that reaction, a tendency to develop in this or that way. This comes out very clearly in regard to cancer: every different inbred strain of mice has a different degree of proneness for a different type of cancer—sometimes 40 per cent., sometimes 80 per cent., in a few cases 100 per cent. Professor Roger Williams of Texas has coined a new word, *propetology*, to denote genetic proneness. A science of propetology is badly needed.

The old-fashioned behaviourists simply denied any influence to genetic factors. For them everything was due to learning; and I am afraid that a number of ethologists and students of behaviour, especially in America, still stick to that point of view. They forget that even the *capacity* to learn, to learn at all, to learn only at a definite stage in development, to learn one kind of thing rather than another, to learn more or less quickly, must have some genetic basis.

One of the most curious discoveries of the past thirty or forty years has been that of the sensory morphisms, where a considerable proportion of the population has a sensory awareness different from that of the "normal" majority. The best known case is a taste-morphism. Phenylthiocarbamide (PTC), tastes very bitter to the majority of human beings, but a minority of about 25 per cent., varying somewhat in different ethnic groups, cannot taste it at all except in exceedingly high concentrations. As R. A. Fisher pointed out in 1930 in his great book, *The Genetical Basis of Natural Selection*, two sharply contrasted genetic characters like this cannot coexist indefinitely in a population unless there is a balance of biological advantage and disadvantage between them. Thus whenever we find such balanced polymorphisms, or *morphisms* as they are more simply called, we know that there must be some selective balance involved. Quite recently it has been shown that PTC taste-morphism is correlated with thyroid function; here we begin to get some inkling of what advantage or disadvantage there may be.

Years ago, Fisher, Ford and I tested all the captive chimpanzees in England for PTC sensitivity. We found, to our delight, that within the limits of statistical error they had the same proportion of non-tasters as human beings. People asked, "How did you find out?" Actually it was quite simple; we offered them a sugar solution containing PTC. If they were non-tasters, they drank it up and put the cup out for more; if they were tasters, they spat it in our faces: it was an all-or-nothing reaction. The fact that both chimpanzees and man react alike means that this balanced morphism, like that of some blood-group systems, must have been in existence in the higher primates since the Pliocene period at least.

There are a number of these sensory morphisms in man. There is a sex-linked morphism with regard to the smell of hydrogen cyanide, HCN; about 18 per cent. of males are insensitive to it, which can be dangerous in a chemical works or laboratory. There is another smell-blindness with regard to the scent of Freesias. I personally am one of the considerable minority of human beings unable to smell Freesias; I can smell any other flower, but am absolutely insensitive to the particular smell of even the most fragrant Freesias. There are visual morphisms; the best known is red-green colour-blindness, which is also sex-linked. Another appears to be myopia. I remember years ago discussing with Professor H. J. Muller the puzzling fact of the considerable incidence of apparently genetic myopia in modern populations. However, he pointed out that during a considerable period of human history, from the time when people began doing fine, close work and up to the period when spectacles were invented, myopia would confer certain advantages. The short-sighted man would not only be employed on well-paid work, but would usually not be sent to war, so that there was less likelihood of his being killed. This would balance the obvious disadvantage of myopia in other aspects of life.

There are some very interesting biological problems concerning sensitivity to pain. Some genetically abnormal human beings are apparently insensitive to pain altogether and may incur terrible injuries because damaging agencies do not hurt them; but these are very rare. On the other hand, giraffes have mouth-cavities and tongues which appear to be surprisingly insensitive to pain. I always thought that they used their beautiful long tongues to strip the leaves off the extremely thorny acacia trees on which they often feed without getting pricked; but apparently the facts are not so simple. Recently in the London Zoo, giraffes have been tested with spiny hawthorn branches: they accept them and chew them just as readily as soft foliage. This surprising fact is worth further investigation.

At the other end of the psychometabolic scale from sensation, we have problems like schizophrenia. Apparently this too must involve a balanced morphism. First, in all countries and races there are about 1 per cent. of schizophrenic people; secondly, the disease appears to have a strong genetic basis; and thirdly, as already mentioned, genetic theory makes it plain that a clearly disadvantageous genetic character like this cannot persist in this frequency in a population unless it is balanced by some compensating advantage. In this case it appears that the advantage is that schizophrenic individuals are considerably less sensitive than normal persons to histamine, are much less prone to suffer from operative and wound shock, and do not suffer nearly so much from various allergies. Meanwhile, there are indications that some chemical substance, apparently something like adrenochrome or adrenolutin, is the genetically-determined basis for schizophrenia and in any case there is a chromatographically detectable so-called "mauve-factor" in the urine of schizophrenics.

This biochemical abnormality presumably causes the abnormality of perception found in schizophrenics. The way the schizophrenic psychometabolizes his sensory experience and relates his sensations to build meaningful perceptions, is disordered. Accordingly he is subject to disorders of sensation and of all sorts of perception, including disorders of perception of time and space, and of association. Apparently schizophrenic individuals show much less consistency in association tests than do normal people. The schizophrenic's world is neither consistently meaningful nor stable: this naturally puts him out of joint with his fellow human beings and makes communication with them difficult and frustrating, so that he retires much more into his own private world.

Hallucinogens like mescaline, lysergic acid and psilocybin (from a Mexican fungus) appear to exert similar dislocating effects on perception, even in incredibly low doses. In addition, they can produce totally new types of experience: some of their effects can elicit something quite new from the human mind. They may have unpleasant effects if the subject is in a wrong psychological state, and exceedingly pleasant and rewarding effects if he is in a right one. But in either case they may reveal possibilities of experience which the subject did not know existed at all. For this reason the term *psychedelic*, or mind-revealing, has been suggested for this type of psychotropic drug. In some ways their effect closely resembles a very brief but acute schizophrenia: perception is disordered in a way very like that seen in schizophrenic patients.

In psychedelic drugs we have a remarkable opportunity for interesting research. Nobody, so far as I know, has done any work on their effects on different types of psychologically normal people—people of high and low I.Q., of different somatotypes, of different affective dispositions, on verbalizers and visualizers. This would be of extraordinary interest: we might find out not merely how to cure some defect, but how to promote creativity by enhancing the creative imagination.

Another problem is to discover whether psychedelics modify or enhance dreaming? The study of dreaming has received a great impetus since the recent discovery that dreaming is necessary for good mental health. If people are prevented from dreaming night after night, their mental health begins to suffer. Dreaming, it seems, provides a satisfactory way of psychometabolizing various facts and experiences that have proved resistant to the integrating efforts of our waking psychometabolic activity. Unconscious mechanisms take revenge and provide an outlet in dreams.

Early detection is another facet of the schizophrenia problem. Here too, study and research are obviously needed. Granted that there is a genetic proneness to schizophrenia, it should be possible in many cases to detect its symptoms in quite early stages of life, and then use appropriate methods of education and training to prevent the disorder from developing to full manifestation.

Indeed, the subject of education in general clearly needs

68

overhauling. Today, we have hardly begun to think of how to educate the organism as a whole—the mind-body, the unitary psychophysiological mechanism which we call the human child. We confine ourselves almost entirely to mental education through verbal means, with the crude physical education of games and physical training added as something quite separate. As my brother Aldous has stressed, we need non-verbal education as well, and education of the entire mind-body instead of "mind" and "body" separately.

It is not only in regard to schizophrenia that we are confronted with situations which demand immediate remedial measures, but later find ourselves impelled to adopt a preventive or a constructive attitude. Medical history is largely the story of people trying to cope with disease, then attempting to prevent disease from arising, and finally turning their knowledge to good account in the promotion of positive health. The same is true of the psychological approach. The psychiatrist starts with the mentally diseased person, tries to cure him, or at least to prevent his disease developing further, but in the course of this remedial process, he acquires knowledge which can be of extreme importance in building up a more fruitful normal personality. However, to achieve this, a new approach is needed. Psychiatry usually attempts to analyse the causes of the diseased condition and discover its origins. The very term *psychoanalysis* commits the Freudian practitioner to this approach. This is important, but is certainly not sufficient. All important biological phenomena are constructive processes, whose end-results are biologically more significant than their origins. Accordingly, we must study the whole process, its end-results as well as its origins, its total pattern as well as its elements.

In psychology, pure and applied, medical and educational, our main aim should be to discover how to regulate the processes of psychometabolism so that they integrate experience in a more effective and less wasteful way and produce a more fruitful end-organization. For instance, I am sure that a study of the origin and strengthening of emotional bonds will repay a great deal of effort. Let me take John Bowlby's work

69

as an example. He studied the development of children who had been deprived of maternal care (including care by a mother substitute) during a critical period of early life, and therefore were unable to form the primary affectional bond between infant and parent. Such children proved incapable of forming further emotional bonds and of developing a normal affectional and moral organization. This whole problem of building up affectional bonds, whether between members of a family or a social group, is fundamental for human life.

The overriding psychometabolic problem, of course, is how the developing human being can integrate his interior life, whether by reconciling emotional or intellectual conflict in a higher synthesis, or by reconciling diversity in a more embracing unity. Let me take the creative arts as an example. Thus the poet must reconcile diverse and even conflicting meanings in a single work of art, and indeed, must employ multivalent or multi-significant words and phrases in the process. Good poems and paintings are among the highest products of man's psychometabolic activities. Milton's line "Then feed on thoughts that voluntary move Harmonious numbers..." beautifully expresses this psychometabolic concept of artistic creation, while Lowes' celebrated critical study, *The Road to Xanadu*, shows how Coleridge psychometabolized the raw materials of his personal experience, his reading and his conversations and discussions, and was able to integrate them into a single poem, "Kubla Khan", a work of art with an amazing emotional impact.

As an example of the emotional impact exercised by great art, let me recall a story of Bertrand Russell. When he was an undergraduate at Cambridge, he and a friend were going up his staircase in College and the friend quoted Blake's famous poem, "Tiger! Tiger! burning bright" ... Bertrand Russell had never heard or read this before, and was so overcome that he had to lean against the wall to prevent himself falling.

On the other hand, we all know that many poems and works of art fail sadly to achieve this desirable result. The way in which an operatively effective unitary pattern of intellectual, emotional and moral elements can be built up certainly deserves study, not merely in art, but also in morality, religion and love.

Mysticism is another psychometabolic activity which needs much further research. A really scientific study of the great mystics of the past, of their modern successors, of yoga and other similar movements, undoubtedly would be of great value. The scientist need not, indeed must not, accept at their face value the claims of mysticism, for instance, of achieving union with God or the Absolute. But some mystics have certainly obtained results of great value and importance: they have been able to achieve an interior state of peace and strength which combines profound tranquillity with high psychological energy.

There is also the still much neglected subject of hypnotism and hypnosis, with all its implications. One of the darker chapters in the history of science and medicine is the way in which the pioneer hypnotists were attacked and often hounded out of the medical profession. Even today, there is still clearly a great deal to be discovered in this strange and exciting subject.

The field of psychiatry and psychology today is nothing less than the comprehensive study of hypnosis, drugs, education, mysticism and the subconscious, of mental disease and mental health, with a view to a better integration of all the psychological forces operating in man's life.[1]

[1] Since writing the above, I have seen a film by Drs Osmond, Hoffer and Fogel demonstrating the extraordinary psychometabolic effects on personality of a hypnotic suggestion of altered perception, whether of time, sound, colour or smell, operating during the post-hynotic period. Thus when perceptual time was slowed down, the subject became lethargic and depressed, culminating in catatonia with waxy flexibility; while speeding up induced quicker movement, speech and thought, and increasing euphoria, culminating in dangerous manic outbursts. Comparable transformations of temperament and attitude were found with other modes of perception. In all cases, personal identity was apparently destroyed: the subjects forgot their names and addresses and whether they were married or not.

THE HUMANIST FRAME

MAN is embarked on the psychosocial stage of evolution. Major advance in that stage of the evolutionary process involves radical change in the dominant idea-systems. It is marked by the passage from an old to a new general organization of thought and belief; and the new pattern of thinking and attitude is necessitated by the increase of knowledge, demanding to be organized in new and more comprehensive ways, and by the failure of older ideas which attempted to organize beliefs round a core of ignorance.

General idea-systems are always concerned, consciously or unconsciously, with beliefs about human destiny, and always influence men's general attitude to life and approach to practical affairs. People brought up in different idea-systems find it difficult to understand each other's approaches and attitudes. Modern industrial man finds it hard to understand tribal peoples, whose idea-systems are organized round the concept of magic power; and equally difficult to understand medieval Western man, whose idea-system was centred round the concept of a central earth, created and ruled by an omnipotent, omniscient and omnibenevolent supernatural Being.[1]

The present uneasy age of disillusion after two Great Wars has witnessed a widespread breakdown of traditional beliefs, but also a growing realization that a purely materialistic outlook cannot provide an adequate basis for human life. It has also witnessed a fantastic growth of knowledge—about the material universe, about life and mind, about human nature and human societies, about art and history and religion; but large chunks of this new knowledge are lying around unused, not worked up or integrated into fruitful

---

[1] A salutary exercise for an inhabitant of the mid-twentieth century is to try to give sympathetic consideration to the arguments about angels in St Thomas Aquinas's *Summa Theologiae*, or to attempt to understand the ideological basis of many of the practices described in Frazer's *Golden Bough*.

72

concepts and principles, not brought into relevance to human life and its problems.

Meanwhile an increasing number of people are coming to feel that man must rely only on himself in coping with the business of living and the problem of destiny, but feel increasingly sceptical about the possibility of his achieving this at all adequately.

If the situation is not to lead to chaos, despair or escapism, man must reunify his life within the framework of a satisfactory idea-system. To achieve this, he needs to survey the resources available to him, both in the outer world and within himself, to define his aims and chart his position, and to plan the outline of his future course. He needs to use his best efforts of knowledge and imagination to build a system of thought and belief which will provide both a supporting framework for his present existence, an ultimate or ideal goal for his future development as a species, and a guide and directive for practical action and planning.

This new idea-system, whose birth we of the mid-twentieth century are witnessing, I shall simply call *Humanism*, because it can only be based on our understanding of man and his relations with the rest of his environment. It must be focused on man as an organism, though one with unique properties. It must be organized round the facts and ideas of evolution, taking account of the discovery that man is part of a comprehensive evolutionary process, and cannot avoid playing a decisive role in it.

Such an Evolutionary Humanism is necessarily unitary instead of dualistic, affirming the unity of mind and body; universal instead of particularist, affirming the continuity of man with the rest of life, and of life with the rest of the universe; naturalistic instead of supernaturalist, affirming the unity of the spiritual and the material; and global instead of divisive, affirming the unity of all mankind. *Nihil humanum a me alienum puto* is the Humanist's motto. Humanism thinks in terms of directional process instead of in those of static mechanism, in terms of quality and diversity as well as quantity and unity. It will have nothing to do with Absolutes, including absolute truth, absolute morality, absolute perfec-

73

tion and absolute authority, but insists that we can find standards to which our actions and our aims can properly be related. It affirms that knowledge and understanding can be increased, that conduct and social organization can be improved, and that more desirable directions for individual and social development can be found. As the overriding aim of evolving man, it is driven to reject power, or mere numbers of people, or efficiency, or material exploitation, and to envisage greater fulfilment and fuller achievement as his true goal.

Most important of all, it brings together the scattered and largely unutilized resources of our knowledge, and orders them to provide a new vision of human destiny, illuminating its every aspect, from the broad and enduring sweep of cosmic process to present-day polities, from the planetary web of world ecology to the individual lives entangled in it, from the dim roots of man's past to the dawning possibilities of his far future.

This new vision is inevitably an evolutionary one. At the University of Chicago's Centennial Celebration of Darwin's launching of the theory of evolution in 1859, I was honoured by being asked to give the Commemoration Address. To give some idea of this new vision, I cannot do better than quote from it:

*   *   *

This centennial celebration is one of the first occasions on which it has been frankly faced that all aspects of reality are subject to evolution, from atoms and stars to fish and flowers, from fish and flowers to human societies and values—indeed that all reality is a single process of evolution. And ours is the first period in which we have acquired sufficient knowledge to begin to see the outline of this vast process as a whole.

Our evolutionary vision now includes the discovery that biological advance exists, and that it takes place in a series of steps or grades, each grade occupied by a successful group of animals or plants, each group sprung from a pre-existing one and characterized by a new and improved pattern of organization.

74

Improved organization gives biological advantage. Accordingly the new type becomes a successful or *dominant* group. It spreads and multiplies and differentiates into a multiplicity of branches. This new biological success is usually achieved at the biological expense of the older dominant group from which it sprang, or whose place it has usurped. Thus the rise of the placental mammals was correlated with the decline of the terrestrial reptiles, and the birds replaced the pterosaurs as dominant in the air.

Occasionally, however, when the breakthrough to a new type of organization is also a breakthrough into a wholly new environment, the new type may not come into competition with the old, and both may continue to co-exist in full flourishment. Thus the evolution of land vertebrates in no way interfered with the continued success of the sea's dominant group, the teleost bony fish.

The successive patterns of successful organization are stable patterns: they exemplify continuity, and tend to persist over long periods. Reptiles have remained reptiles for a quarter of a billion years: tortoises, snakes, lizards and crocodiles are all still recognizably reptilian, all variations of one organizational theme.

It is difficult for life to transcend this stability and achieve a new successful organization. That is why breakthroughs to new dominant types are so rare—and also so important. The reptilian type radiated out into well over a dozen important groups or Orders: but all of them remained within the reptilian framework except two, which broke through to the new and wonderfully successful patterns of bird and mammal.

In the early stages, a new group, however successful it will eventually become, is few and feeble and shows no signs of the success it may eventually achieve. Its breakthrough is not an instantaneous matter, but has to be implemented by a series of improvements which eventually become welded into the new stabilized organization.

With mammals there was first hair, then milk, then partial and later full temperature-regulation, then brief and finally prolonged internal development, with evolution of a placenta. Mammals of a small and insignificant sort had existed and evolved for a hundred million years or so before they achieved the full breakthrough to their explosive dominance in the Cenozoic.

Something very similar occurred during our own break-through from mammalian to psychosocial organization. Our prehuman ape ancestors were never particularly successful or abundant. For their transformation into man a series of steps were needed. Descent from the trees; erect posture; some enlargement of brain; more carnivorous habits; the use and then the making of tools; further enlargement of brain; the discovery of fire; true speech and language; elaboration of tools and rituals. These steps took the better part of half a million years; it was not until less than a hundred thousand years ago that man could begin to deserve the title of dominant type, and not till less than ten thousand years ago that he became fully dominant.

After man's emergence as truly man, this same sort of thing continued to happen, but with an important difference. Man's evolution is not biological but psychosocial: it operates by the mechanism of cultural tradition, which involves the cumulative self-reproduction and self-variation of mental activities and their products. Accordingly, major steps in the human phase of evolution are achieved by break-throughs to new dominant patterns of mental organization, of knowledge, ideas and beliefs—ideological instead of physiological or biological organization.

There is thus a succession of successful idea-systems instead of a succession of successful bodily organizations. Each new successful idea-system spreads and dominates some important sector of the world, until it is superseded by a rival system, or itself gives birth to its successor by a break-through to a new organized system of thought and belief. We need only think of the magic pattern of tribal thought, the god-centred medieval pattern organized round the concept of divine authority and revelation, and the rise in the last three centuries of the science-centred pattern, organized round the concept of human progress, but progress somehow under the control of supernatural authority. In 1859 Darwin opened the door to a new pattern of ideological organization—the evolution-centred organization of thought and belief.

Through the telescope of our scientific imagination we can discern the existence of this new and improved ideological organization, albeit in embryonic form; but many of its details are not yet clear, and we can also see that the upward steps needed to reach its full development are many and hard to take.

Let me change the metaphor. To all save those who deliberately shut or averted their eyes, or were not allowed by their pastors or masters to look, it was at once clear that the fact and concept of evolution was bound to act as the central germ or living template of a new dominant thought-organization. And in the century since the *Origin of Species* there have been many attempts to understand the implications of evolution in many fields, from the affairs of the stellar universe to the affairs of men, and a number of preliminary and largely premature efforts to integrate the facts of evolution and our knowledge of its processes into the overall organization of our general thought.

All dominant thought-organizations are concerned with the ultimate as well as with the immediate problems of existence: or, I should rather say, with the most ultimate problems that the thought of the time is capable of envisaging or even formulating. They are all concerned with giving some interpretation of man, of the world which he is to live in, and of his place and role in that world—in other words some comprehensible picture of human destiny and significance.

The broad outlines of the new evolutionary picture of ultimates are beginning to be clearly visible. Man's destiny is to be the sole agent for the future evolution of this planet. He is the highest dominant type to be produced by over two and a half billion years of the slow biological improvement effected by the blind opportunistic workings of natural selection; if he does not destroy himself, he has at least an equal stretch of evolutionary time before him to exercise his agency.

During the later part of biological evolution, mind—our

word for the mental activities and properties of organisms—
emerged with greater clarity and intensity, and came to play
a more important role in the individual lives of animals.
Eventually it broke through to become the foundation and
the main source of further evolution, though the essential
character of evolution now became cultural instead of genetic
or biological. It was to this breakthrough, brought about by
the automatic mechanism of natural selection and not by any
conscious effort on his own part, that man owes his dominant
evolutionary position.

Man therefore is of immense significance. He has been
ousted from his self-imagined centrality in the universe to an
infinitesimal location in a peripheral position in one of a
million of galaxies. Nor, it would appear, is he likely to be
unique as a sentient being. On the other hand, the evolution
of mind or sentiency is an extremely rare event in the vast
meaninglessness of the insentient universe, and man's
particular brand of sentiency may well be unique. But in any
case he is highly significant. He is a reminder of the existence,
here and there, in the quantitative vastness of cosmic matter
and its energy-equivalents, of a trend towards mind, with its
accompaniment of quality and richness of existence; and,
what is more, a proof of the importance of mind and quality
in the all-embracing evolutionary process.

It is only through possessing a mind that he has become
the dominant portion of this planet and the agent responsible
for its future evolution; and it will only be by the right use of
that mind that he will be able to exercise that responsibility
rightly. He could all too readily be a failure in the job; he will
only succeed if he faces it consciously and if he uses all his
mental resources—knowledge and reason, imagination and
sensitivity, capacities for wonder and love, for comprehension
and compassion, for spiritual aspiration and moral effort.

And he must face it unaided by outside help. In the
evolutionary pattern of thought there is no longer either need
or room for the supernatural. The earth was not created: it
evolved. So did all the animals and plants that inhabit it,
including our human selves, mind and soul as well as brain
and body. So did religion. Religions are organs of psycho-

social man concerned with human destiny and with experiences of sacredness and transcendence. In their evolution, some (but by no means all) have given birth to the concept of gods as supernatural beings endowed with mental and spiritual properties and capable of intervening in the affairs of nature, including man. These theistic religions are organizations of human thought in its interaction with the puzzling, complex world with which it has to contend—the outer world of nature and the inner world of man's own nature. In this, they resemble other early organizations of human thought confronted with nature, like the doctrine of the Four Elements, earth, air, fire and water, or the Eastern concept of rebirth and reincarnation. Like these, they are destined to disappear in competition with other, truer, and more embracing thought-organizations which are handling the same range of raw or processed experience.

Evolutionary man can no longer take refuge from his loneliness by creeping for shelter into the arms of a divinized father-figure whom he has himself created, nor escape from the responsibility of making decisions by sheltering under the umbrella of Divine Authority, nor absolve himself from the hard task of meeting his present problems and planning his future by relying on the will of an omniscient but unfortunately inscrutable Providence. On the other hand, his loneliness is only apparent. He is not alone as a type. Thanks to the astronomers, he now knows that he is one among the many organisms that bear witness to the trend towards sentience, mind and richness of being, operating so widely but so sparsely in the cosmos. More immediately important, thanks to Darwin, he now knows that he is not an isolated phenomenon, cut off from the rest of nature by his uniqueness. Not only is he made of the same matter and operated by the same energy as all the rest of the cosmos, but for all his distinctiveness, he is linked by genetic continuity with all the other living inhabitants of his planet. Animals, plants and micro-organisms, they are all his cousins or remoter kin, all parts of one single branching and evolving flow of metabolizing protoplasm.

Nor is he individually alone in his thinking. He exists and

has his being in the intangible sea of thought which Teilhard de Chardin has christened the noösphere, in the same sort of way that fish exist and have their being in the material sea of water which the geographers include in the term hydrosphere. Floating in this noösphere there are, for his taking, the daring speculations and aspiring ideals of man long dead, the organized knowledge of science, the hoary wisdom of the ancients, the creative imaginings of all the world's poets and artists. And in his own nature there are, waiting to be called upon, an array of potential helpers—all the possibilities of wonder and knowledge, of delight and reverence, of creative belief and moral purpose, of passionate effort and embracing love.

Turning the eye of an evolutionary biologist on the situation, I would compare the present stage of evolving man to the geological moment, some three hundred million years ago, when our amphibian ancestors were just establishing themselves out of the world of water. They had created a bridgehead into a wholly new environment. No longer buoyed up by water, they had to learn how to support their own weight; debarred from swimming with their muscular tail, they had to learn to crawl with clumsy limbs. The newly discovered realm of air gave them direct access to the oxygen they needed to breathe, but it also threatened their moist bodies with desiccation. And though they managed to make do on land during their adult existence, they found themselves still compulsorily fishy during the early stages of their lives.

On the other hand, they had emerged into completely new freedom. As fish, they had been confined below a bounding surface. Now the air above them expanded out into the infinity of space. Now they were free of the banquet of small creatures prepared by the previous hundred million years of life's terrestrial evolution. The earth's land surface provided a greater variety of opportunity than did its waters, and also a much greater range of challenge to evolving life. Could the early Stegocephalians have been gifted with imagination, they might have seen before them the possibility of walking, running, perhaps even flying over the earth; the probability of their descendants escaping from bondage to winter cold by regulating their temperature, escaping from bondage to the waters by constructing private ponds for their early development; the inevitability of an upsurge of their dim minds to new levels of clarity and performance. But meanwhile they would see themselves tied to an ambiguous existence, neither one thing nor the other, on the narrow moist margin between water and air. They could have seen the promised land afar off, though but dimly through their bleary newtish eyes. But they would also have seen that, to reach it, they would have to achieve many difficult and arduous transformations of their being and way of life.

So with ourselves. We have only recently emerged from the biological to the psychosocial area of evolution, from the earthly biosphere into the freedom of the noösphere. Do not let us forget how recently: we have been truly men for perhaps a tenth of a million years—one tick of evolution's clock: even as proto-men, we have existed for under one million years—less than a two-thousandth fraction of evolutionary time. No longer supported and steered by a framework of instincts, we try to use our conscious thoughts and purposes as organs of psychosocial locomotion and direction through the tangles of our existence; but so far with only moderate success, and with the production of much evil and horror as well as of some beauty and glory of achievement. We too have colonized only an ambiguous margin between an old bounded environment and the new territories of freedom. Our feet still drag in the biological mud, even when we lift our heads into the conscious air. But unlike those remote ancestors of ours, we can truly see something of the promised land beyond. We can do so with the aid of our new instrument of vision—our rational, knowledge-based imagination. Like the earliest pre-Galilean telescopes, it is still a very primitive instrument, and gives a feeble and often distorted view. But, like the early telescopes, it is capable of immense improvement, and could reveal many secrets of our noöspheric home and destiny.

Meanwhile no mental telescope is needed to see the immediate evolutionary landscape, and the frightening problems which inhabit it. All that is required—but that is plenty—is for us to cease being intellectual and moral ostriches, and take our heads out of the sand of wilful blindness. If we do so, we shall soon see that the alarming problems are two-faced, and are also stimulating challenges.

What are those challenging monsters in our evolutionary path? I would list them as follows. The threat of super-scientific war, nuclear, chemical, and biological; the threat of over-population; the rise and appeal of Communist ideology, especially in the under-privileged sectors of the world's people; the failure to bring China, with nearly a quarter of the world's population, into the world organization of the United Nations; the over-exploitation of natural resources; the erosion of the world's cultural variety; our general pre-occupation with means rather than ends, with technology and quantity rather than creativity and quality; and the Revolution of Expectation, caused by the widening gap between the haves and the have-nots, between the rich and the poor nations. This day of Darwinian celebration is Thanksgiving Day in America. But millions of people now living have little cause to give thanks for anything. When I was in India last spring, a Hindu man was arrested for the murder of his small son. He explained that his life was so miserable that he had killed the boy as a sacrifice to the goddess Kali, in the hope that she would help him in return. That is an extreme case, but let us remember that two-thirds of the world's people are under-privileged—under-fed, under-healthy, under-educated—and that many millions of them live in squalor and suffering. They have little to be thankful for save hope that they will be helped to escape from this misery. If we in the West do not give them aid, they will look to other systems for help—or even turn from hope to destructive despair.

We attempt to deal with these problems piecemeal, often half-heartedly; sometimes, as with population, we refuse to recognize it officially as a World Problem (just as we refuse to recognize Communist China as a World Power). In reality, they are not separate monsters to be dealt with by a series of separate ventures, however heroic or saintly. They are all symptoms of a new evolutionary situation; and this can only be successfully met in the light and with the aid of a new organization of thought and belief, a new dominant pattern of ideas relevant to the new situation.

It is hard to break through the firm framework of an accepted belief-system, and build a new and complex successor, but it is necessary. It is necessary to organize our *ad hoc* ideas and scattered values into a unitive pattern, transcending conflicts and divisions in its unitary web. Only by such a reconciliation of opposites and disparates can our belief-system release us from inner conflicts: only so can we gain that peaceful assurance which will help unlock our energies for development in strenuous practical action.

Our new pattern of thinking will be evolution-centred. It will give us assurance by reminding us of our long evolutionary rise; how this was also, strangely and wonderfully, the rise of mind; and how that rise culminated in the eruption of mind as the dominant factor in evolution and led to our own spectacular but precarious evolutionary success. It will give us hope by pointing to the aeons of evolutionary time that lie ahead of our species if it does not destroy itself or nullify its own opportunities; by recalling how the increase of man's understanding and the improved organization of knowledge has in fact enabled him to make a whole series of advances, such as control of infectious disease or efficiency of tele-communication, and to transcend a whole set of apparently unbridgeable oppositions, like the conflict between Islam and Christendom, or that between the seven kingdoms of the Heptarchy; and by reminding us of the vast stores of human effectiveness—of intelligence, imagination, co-operative goodwill—which still remain untapped.

Our new organization of thought—belief-system, framework of values, ideology, call it what you will—must grow and be developed in the light of our new evolutionary vision. So, in the first place, it must of course itself be evolutionary: that is to say, it must help us to think in terms of an overriding process of change, development, and possible improvement, to have our eyes on the future rather than on the past, to find support in the growing, spreading, upreaching body

83

of our knowledge, instead of in the rigid frame of fixed dogma or ancient authority. Equally, of course, the evolutionary outlook must be scientific, not in the sense that it rejects or neglects other human activities, but in believing in the value of the scientific method for eliciting knowledge from ignorance and truth from error, and in basing itself on the firm ground of scientifically established knowledge. Unlike most theologies, it accepts the inevitability and indeed the desirability of change, and advances by welcoming new discovery even when it conflicts with old ways of thinking.

The only way in which the present split between religion and science could be mended would be through the acceptance by science of the fact and value of religion as an organ of evolving man, and the acceptance by religion that religions must evolve if they are not to become extinct, or at best turn into outdated living fossils struggling to survive in a new and alien environment.

Next, the evolutionary outlook must be global. Man is strong and successful in so far as he operates in interthinking groups, which are able to pool their knowledge and beliefs. To have any success in fulfilling his destiny as the controller or agent of future evolution on earth, he must become one single inter-thinking group, with one general framework of ideas: otherwise his mental energies will be dissipated in ideological conflict. Science gives us a foretaste of what could be. It is already global, with scientists of every nation contributing to its advance: and because it is global, it is advancing fast. In every field we must aim to transcend nationalism: and the first step towards this is to think globally —how could this or that task be achieved by international co-operation rather than by separate action?

But our thinking must also be concerned with the individual. The well-developed well-patterned individual human being is, in a strictly scientific sense, the highest phenomenon of which we have any knowledge; and the variety of individual personalities is the world's highest richness.

In the light of the evolutionary vision the individual need not feel just a meaningless cog in the social machine, nor merely the helpless prey and sport of vast impersonal forces. He can do something to develop his own personality, to discover his own talents and possibilities, to interact personally and fruitfully with other individuals, to discover something of his own significance. If so, in his own person he is realizing an important quantum of evolutionary possibility; he is contributing his own personal quality to the fulfilment of human destiny; and he has assurance of his own significance in the vaster and more enduring whole of which he is a part.

I spoke of quality. This must be the dominant concept of our belief-system—quality and richness as against quantity and uniformity. Though our new idea-pattern must be unitary, it need not and should not be cramping, or impose a drab or boring cultural uniformity. An organized system, whether of thought, expression, social life or anything else, has some degree both of unity and richness. Cultural variety, both in the world as a whole and within its separate countries, is the spice of life; and individual variety is the necessary basis for personal fulfilment as well as for social efficiency. Yet both are being threatened and indeed eroded away by mass-production, mass-communications, mass-education, mass-conformity, and all the other forces making for uniformization—an ugly word for an ugly thing! We have to work hard to preserve and foster them.

Population is people in the mass; and it is in regard to population that the most drastic reversal or reorientation of our thinking has become necessary. The unprecedented population-explosion of the last half-century has strikingly exemplified the Marxist principle of the passage of quantity into quality. Mere increase in quantity of people is increasingly affecting the quality of their lives and their future, and affecting it almost wholly for the worse.

Population-increase is already destroying or eroding many of the world's resources, both those for material subsistence and those—equally essential but often neglected—for human enjoyment and fulfilment. Early in man's history the injunction to increase and multiply was right. Today it is wrong, and to obey it will be disastrous. The world has to achieve the difficult task of reversing the direction of its thought about population. It has to begin thinking that our aim should be not increase but decrease—immediate decrease in the rate of population-growth; and in the long run, decrease in the absolute number of people in the world, including our own countries.

We must make the same reversal of ideas about economics. At the moment our Western economic system (which is steadily invading new regions) is based on expanding production for profit; and production for profit is based on expanding consumption. As one writer has put it, the American economy depends on persuading more people to believe that they want to consume more products. This is leading to gross over-exploitation of resources that ought to be conserved, to excessive advertising, to the dissipation of talent and energy into unproductive channels, and to a diversion of the economy as a whole away from its true functions.

But, like the population-explosion, this consumption-explosion cannot continue much longer: it is an inherently self-defeating process. Sooner rather than later we shall be forced to get away from a system based on artificially increasing the number of human wants, and set about constructing one aimed at the qualitative satisfaction of real human needs, spiritual and mental as well as material and physiological. This means abandoning the pernicious habit of evaluating every human project solely in terms of its utility—by which the evaluators mean solely its material utility, and especially its utility in making a profit for somebody. Once we truly believe that man's destiny is to make possible greater fulfilment for more human beings and fuller achievement by human societies, utility in the customary sense becomes subordinate. Quantity of material production is, of course, necessary as the basis for the satisfaction of elementary human needs—but only up to a certain degree. More than a certain number of calories or cocktails or TV sets or washing machines per person is not merely unnecessary but bad. Quantity of material production can only be a means to a further end, not an end in itself.

The important ends of man's life include the creation and

86

enjoyment of beauty, both natural and man-made; increased comprehension and a more assured sense of significance; the preservation of all sources of pure wonder and delight, like fine scenery, wild animals in freedom, or unspoiled nature; the attainment of inner peace and harmony; the feeling of active participation in embracing and enduring projects, including the cosmic project of evolution. It is through such things that individuals attain greater fulfilment. As for nations and societies, they are remembered not for their wealth or comforts or technologies, but for their great buildings and works of art, their achievements in science or law or political philosophy, their success in liberating human life from the shackles of fear and ignorance.

Although it is to his mind that man owes both his present dominant position in evolution, and any advances he may have made during his tenure of that position, he is still strangely ignorant and even superstitious about it.[1] The exploration of the mind has barely begun. It must be one of the main tasks of the coming era, just as was the exploration of the world's surface a few centuries ago. Psychological exploration will doubtless reveal as many surprises as did geographical exploration, and will make available to our descendants all kinds of new possibilities of full and richer living.

Finally, the evolutionary vision is enabling us to discern, however incompletely, the lineaments of the new religion that we can be sure will arise to serve the needs of the coming era. Just as stomachs are bodily organs concerned with digestion, and involving the biochemical activity of special juices, so are religions psychosocial organs concerned with the problems of human destiny, and involving the emotion of sacredness and the sense of right and wrong. Religion of some sort is probably necessary. But it is not necessarily a good thing. It was not a good thing when the Hindu I read

---

[1] The pseudo-scientific behaviourist superstition consists in denying it any effective existence beyond that of a pale ghost; and in dismissing it and its products as outside the range of scientific investigation. The philosophical idealist superstition, on the contrary, denies the effective existence of anything else.

about last spring killed his son as a religious sacrifice. It is not a good thing that religious pressure has made it illegal to teach evolution in Tennessee, because it conflicts with Fundamentalist beliefs. It is not a good thing that in Connecticut and Massachusetts women should be subject to grievous suffering because Roman Catholic pressure refuses to allow even doctors to give information on birth-control even to non-Catholics. It was not a good thing for Christians to persecute and even burn heretics; it is not a good thing when Communism, in its dogmatic-religious aspect, persecutes and even executes deviationists.

The emergent religion of the near future could be a good thing. It will believe in knowledge. It will be able to take advantage of the vast amount of new knowledge produced by the knowledge-explosion of the last few centuries in constructing what we may call its theology—the framework of facts and ideas which provide it with intellectual support: it should be able, with our increased knowledge of mind, to define man's sense of right and wrong more clearly, so as to provide a better moral support, and to focus the feeling of sacredness on fitter objects. Instead of worshipping supernatural rulers, it will sanctify the higher manifestations of human nature, in art and love, in intellectual comprehension and aspiring adoration, and will emphasize the fuller realization of life's possibilities as a sacred trust.

Thus the evolutionary vision, first opened up for us by Charles Darwin a century back, illuminates our existence in a simple but almost overwhelming way. It exemplifies the truth that truth is great and will prevail, and the greater truth that truth will set us free. Evolutionary truth frees us from subservient fear of the unknown and supernatural, and exhorts us to face this new freedom with courage tempered with wisdom, and hope tempered with knowledge. It shows us our destiny and our duty. It shows us mind enthroned above matter, quantity subordinate to quality. It gives our anxious minds support by revealing the incredible possibilities that have already been realized in evolution's past; and, by pointing to the hidden treasure of fresh possibilities that still remain to be realized, it gives us a potent incentive for fulfilling our evolutionary role in the long future of our planet.

*   *   *

I must now attempt, however inadequately, to outline what I conceive to be the Humanist view of the three great activities of man in which he transcends the material business of making a living—art, science and religion.

I use these terms broadly—art to cover all organized expression of experience in significant and aesthetically effective form, science in the continental sense of all organized knowledge and learning, and religion as including all systems of belief and morality primarily concerned with the problem of destiny.

The three types of activity overlap and interlock; but they are essentially distinct, for they perform different psycho-social functions.

Art is almost exclusively a human characteristic. Almost, but not quite. Male bowerbirds show marked aesthetic preferences (for different colours and different kinds of objects for their bowers), and some individuals even paint the lower part of their bowers with pigments; while chimpanzees (and just possibly elephants) provided with the requisite materials will enjoy producing paintings which show an elementary sense of design. But such instances are rare among animals; while in man, art in some form is a universal phenomenon, playing an important role in all types of human society and at all levels of human development.

I leave to the philosophers and aestheticians the job of defining Art: as a Humanist I am concerned with the concrete question of the function of art, or, still more concretely, with what the various arts do in various human societies. First of all we must give ourselves the semantic reminder that there is no such *thing* as art. Art is not an entity, any more than life is an entity. It is a word, a general term conveniently but often loosely used to cover a certain rather wide-ranging type of human activity and its products. It is impossible to delimit either the type or its range with precision. Here I shall use *art* to cover the effective organiza-

89

tion of experience into integrated forms which are emotionally significant and aesthetically satisfying. This includes
some of the practice and some of the products of activities
like painting and sculpture, literature and drama, dance and
ritual celebration, music and architecture. But, of course, in
the spectrum of all these activities art slips over into non-art.
Literature grades into straightforward information and into
propaganda, visual art into decoration on the one hand and
advertisement on the other, drama into mere entertainment,
ritual into meaningless formalism, dance into pastime, architecture into utilitarian and sometimes extremely unaesthetic
building. And also, of course, every art has its own spectrum,
from bad to good: it is impossible to have good art without
some bad and much indifferent art.

The essential distinctness of art, I would say, is that it
provides a qualitative enrichment of life, by creating a
diversity of new experience. For one thing, art can tap
emotional resources of human personality which might
otherwise remain unutilized, either individually or socially—
"dark levels of feeling, both conscious and unconscious,
which are a kind of driving power and a determinant of
happiness". It is "a process of extending ourselves, through
our sensibilities and our imagination, to something we have
not reached before. It is a process of discovery about ourselves and about life."[1] Art helps us to assimilate the experience provided by our senses and emotions. It is an essential
part of our psychometabolic system. "Even as we feed our
bodies, so do we need to feed and sustain the imagination":
and art can potently help in this. Imagine a world without
any art: life in it would be intolerable.

But although art is in general a process of differentiation
and proliferates variety, it is in particular always a process of
integration and synthesis; any work of art, however humble,
brings together a number of separate (and sometimes
apparently disparate) elements and moulds them into an
organic unity.

Art can exert the most profound effects on the minds of

[1] The admirable phrases within inverted commas are from a private
document from which I have been permitted to quote.

men. To many people poetry or painting or music have conveyed an overwhelming sense of revelation. I can recall the overwhelming impression made on me by Giotto's great painted epic in the Arena Chapel at Padua; and has not Keats himself recorded in immortal words the profound impact made on him by his first contact with the works of another great poet, Homer? At the play, as Aristotle said, we can be "purged by pity and fear" or gripped by powerful and liberating collective emotion, and many people have found that their first visit to the theatre was also their induction into a new and compelling mode of experience. We are not quite the same after we have read Tolstoy's *War and Peace*. And Beethoven's posthumous quartets can transport us to another world, make us free of another realm of being.

That is the point. Art opens the doors of that other world in which matter and quantity are transcended by mind and quality. Art is sometimes contemptuously dismissed as escapism. But we all need escape. Apart from our modern need to escape from the dullness of routine and from the over-mechanized life of cities, there is the universal and permanent need to escape from the cage of the practical and actual present in which we have of necessity to spend so much of our lives, and above all from the prison of our single and limited selves. The question is, where and how shall we escape? We can escape downwards, through drink or drugs or dissipation: but that is not the best way. Or sideways, through sport or pastime or entertainment: that is within wide limits desirable and indeed necessary. Or we may escape upwards, into a new world (think of Blake's title, *The Mental Traveller*) comprising new countries of life and new levels of being, where we make contact with something more enduring, more satisfying and in a certain true sense higher than is to be found in the world of material needs and everyday routine.

Man by his very nature has the possibility and indeed the necessity of living his life in two worlds at different levels of meaning—the world of matter and mechanical operations, and that of mind and psychological operations—the level of material needs and that of mental satisfactions. And the

91

mental world is in the strict sense of the word transcendent. In it, we manage to escape from the material world and its quantitative exigencies by transcending it in some higher synthesis in which qualitative elements of our being are organized into effective forms. In the light of evolutionary Humanism, man is seen as struggling, consciously or unconsciously, to create more areas of this matter-transcending world of mental operation, and pressing painfully on towards fuller emergence into its satisfying realms.

Operationally, a work of art exerts its effects by conveying multiple meaning in a single synthesis. The meaning is often best conveyed by suggestion rather than by attempts at rigid and accurate affirmation, as Gombrich so well demonstrates in *Art and Illusion*. The suggestion may work on the basis of long-forgotten and even unconsciously assimilated early experience, or on remembered association, or by way of potent symbol or of effective design. The multiplicity of meaning may be conveyed by single elements in the work— think of the multivalency of words or phrases in great poetry like Blake's *Tiger* or Coleridge's *Kubla Khan*, or Traherne's "the corn was orient and immortal wheat", the multivalency of an individual character in a play, like Shakespeare's *Hamlet*, the bringing together of single elements with separate meanings into a multisignificant whole pattern, as in a great pictorial composition like Raphael's *School of Athens*, or a great musical composition like Bach's B Minor Mass, or a great novel like Tolstoy's *War and Peace*.

The artist can utilize intellectual ideas and moral concepts among the raw materials which he organizes, thus transmuting reason and morality into art and giving a further dimension to his work. In painting, we need only think of the conceptual background of Michelangelo's *Creation of Adam*, of the combination of the concepts of maternity and divinity in pictures of the Madonna and Child, and indeed of all effective use of accepted iconography and symbolism. Greek tragedy flowered out of the ground of current ideas and beliefs, and Dante's *Divina Commedia* owes its compelling greatness largely to the strong and beautifully organized intellectual framework on which it is supported.

Inferior artists will be incapable of organizing these non-aesthetic elements into an aesthetic unity, and their work will not rise above the didactic or the propagandist, the moralistic or the merely representational. But the good artist can fuse them into a richer whole in the creative crucible of his imagination.[1]

The cruciform plan of Christian churches, for all its symbolic and ideological significance, has no particular aesthetic merit. But the great cathedral and abbey builders of the Middle Ages utilized it to produce results of intense architectural value, which the world would otherwise never have enjoyed, by organizing glorious patterns of enclosed space in the meeting of transept, nave and choir. Symbols and ideas are not art, but they can support it, can enrich it, and can enlarge its scope. By posing a problem, they can be a stimulating challenge to the artist, whether architect or painter, in the same way as can problems of material or space or site.

The idea that art is in some way equatable with beauty, or is confined to the creation of beauty, is still widespread, though its fallacy has often been demonstrated. What art creates is significance—emotionally and aesthetically effective significance. Beauty is among its significant products, so that art will increase the store of beauty in the world: but beauty is not its sole or even its main product, and there are many other fields in which beauty can be conserved or created for human fulfilment.

Looked at in the long perspective, both art and science can be said to be progressive, though in different ways. While art increases the qualitative richness and the emotional range of human experience and insight, science increases the volume and the depth of knowledge, its operative efficiency and its better organization, and enlarges the area and the grasp of human understanding.

Art results in the creation of individual works, each qualitatively different from the rest, and embodying a particular organization of experience. A particular work of art may

[1] Books like Kenneth Clark's *Looking at Pictures*, London, Murray, 1960, help us in understanding the artist's creative ability.

93

be timeless, in the sense that men can continue to enjoy it in spite of lapse of time and change of circumstance. Science is mainly cumulative and co-operative; the scientist makes his contribution to a growing and enlarging whole, whose unity is more important and more overriding than the unitary quality of the individual contributions which are incorporated into it. But art is not cumulative or co-operative in this way. It is, in evolutionary terms, a cladogenetic or branching process, promoting differentiation and diversity. In its historical development, art plays somewhat the same role as does adaptive radiation and diversification in biological evolution. Both lead to a fuller utilization of the potential resources and opportunities of the environment—in the case of adaptive radiation, of the material resources and opportunities available for organic metabolism in the physical environment; in the case of art, of the emotional aesthetic resources and possibilities available for psychometabolism in the total environment, both physical and cultural. This is art's main evolutionary aspect. But it is progressive or directional in other ways. It is summatively progressive: Malraux has recently emphasized that, today, for the first time in history, the whole sum of past art is available for present enjoyment. It is technically progressive: painters, for instance, either build on past techniques or react against them and search for new ones. And it is essentially progressive, in that with the lapse of time men not only learn to turn new aspects of experience into art (as Renaissance painters did with space and perspective, or as contemporary artists are doing with abstract or action painting), but discover how to organize a greater number of different components, of thought and emotion as well as of technique, into a single work which shall be a significant whole; in other words, how to create higher patterns of aesthetic organization. A symphony is in a certain real sense a higher achievement than a song or a military march, Giotto's Arena Chapel in Padua than a single drawing or than Lascaux, Dante's *Divina Commedia* than a sonnet, Tolstoy's *War and Peace* than a Maupassant short story or even Don Quixote.

Of course, within each type or grade of organization there

94

is every possible range from good to bad, just as there is every possible range from successful to unsuccessful species and lineages within one grade of biological organization like the mammals or the reptiles. But this does not invalidate the fact that some grades of aesthetic organization are truly "higher" than others, just as mammalian organization is in a strictly scientific sense higher than reptilian. In both cases it is clear that real advance is involved.

It is also clear that in some periods, like the Victorian era, official or generally acceptable art tends to become mediocre or insignificant. This in turn generates minor rebellions or reactions, which may remain without much effective issue, like Pre-Raphaelitism in England, or may burst through the traditional crust to initiate a new and fruitful stream of development, like Impressionism in France.

The rebels usually strengthen their psychological position by forming more or less definite labelled groups—the Pre-Raphaelites, the Barbizon School, the Post-Impressionists, the Cubists, the Surrealists. But a leading role will always be played by individual creative artists of genius and determination—Giotto, Michelangelo, Rembrandt; Constable, Turner, Manet; in our own days, Picasso, Klee, Henry Moore.

There can be little dispute that many developments of the arts in Western countries since 1945 reflect or express the nihilism of the post-war period, the fragmentation of its life between frustration and hope, its intellectual chaos and moral disillusionment. (The rise of existentialism was a symptom of the same pathological malaise.) The very different course of events in Communist countries is also an example of the influence of ideas and beliefs and attitudes on art, in this case the influence of a dogmatic ideology, operating by authoritarian methods and sometimes even downright persecution. In both camps the spread of Humanist ideas would tend to heal the split between creative art and its social environment.

Within its major and embracing function, art has a number of applications. Art can play an important educational role both *in* education and *as* education. Most young children, including especially those who are neither mechanically nor intellectually minded, can enjoy themselves, and (what is more important) find themselves more fully through some form of what is rather pompously called "creative expression", particularly perhaps in paint or clay. It can help them to develop their little personalities in a more emotionally integrated way, and can introduce them both to an enlarged experience of the world and to the discovery of new inner possibilities in themselves. With the self-consciousness of adolescence, it is all too easy to cramp their freedom and conventionalize their expression: but with proper methods the arts can continue to play an important role throughout all the years of education.

Although "creative expression" sounds pompous when applied to young children, it is in fact perfectly correct. All art, even the scribblings and daubings of little boys and girls, not merely provides an outlet for self-expression, but is in some sort a creation, a personal integration imposed by the imagination on some fraction of the fleeting flux of experience. As such, it can act both as a liberating and as an integrative force in the developing human creature.

This combined liberation and integration, one may assume, is the chief function of most art practised as a hobby or a relaxation by adults. It may be egotistically slanted towards self-expression to provide a heightened sense of personal significance; but it seems often to be based simply on the felt need to express some powerful subjective experience in the objective form of a poem or a painting. To achieve this (as anyone knows who has been impelled to write poetry in his youth) demands a special creative effort, but the mere process of achievement, however inadequate, also affords a special satisfaction.

Art may also be therapeutically slanted. The practice of some art can keep certain psychological types from becoming neurotic, and can often assist recovery from a neurotic breakdown.

But though the Humanist is interested in these various ways in which what we may call amateur art may help individual development and promote individual fulfilment, his major concern must be with professional art and its psychosocial functions. How and to what extent does art reflect or express a period or a people; how and to what extent does it promote cultural richness and achievement— in a word, what part does it play in psychosocial evolution?

It is clear that within a given region the arts evolve. Consciously or unconsciously, each new period, each new generation, demands change. Sometimes the change consists primarily in the fuller development or even the exaggeration of an existing style or tradition; sometimes it is a reaction against it, with the emergence of radically new approaches, attitudes, and techniques; sometimes a combination of the two. Gombrich gives plenty of examples in his fascinating book.[1]

I would say that the most general answer runs somewhat like this. The individual artist has two main functions—that of creator and that of interpreter. As interpreter, he translates complex and emotion-tinged experience into directly communicable forms and so is able to express what otherwise would remain unexpressible. He bears witness to the variety of the world and its significance, to its wonder and beauty, but also to its horror and nastiness. His witness may be by way of affirmation or by way of protest. But his function, even when he is not conscious of it, is to interpret the world to man, and man to himself. As creator, on the other hand, he provides experiences of stimulus and enjoyment, sometimes enlargements of experience itself (think of Turner or Stravinsky, or, most obviously, of Shakespeare). Art as a collective social activity has the same two main functions. I should say that it *can* have; for in some societies it is neglected or distorted, or even, as in ancient Sparta, rejected. Nor must we forget that architecture is an art, or, to spell the matter out, that art plays an essential role in architecture, layout and planning. Architecture in this extended sense can perform its own particular function in expressing human ideas and aspirations. Good architecture can enrich human

[1] E. H. Gombrich, *Art and Illusion*, London 1960. See also Kenneth Clark, *The Nude*, for a penetrating study of the changing treatment of a particular subject-matter.

life, especially urban life, while bad architecture can impoverish it, as is all too obvious in the many ugly towns and drab city fringes and subtopian sprawls of our age. To the Humanist, the importance of architecture's social function is obvious: the problem is to persuade officials and taxpayers to recognize its importance in practice.

We must also remember that art spills over into design. Art is as essential for the design of pots and pans and pillar-boxes, of pencils, porcelain and posters, as it is for the quality of paintings or sculptures. Indeed some pots are better works of art than many paintings. And though design has a humbler role than the fine arts, it social function is just as indispensable.

In the fulfilment society envisaged by Humanism, art would be assigned a large role—to beautify the public sector, to bear witness to the richness of existence, to affirm values in concrete effective form, to provide achievements of which human societies can be proud and through which mankind can find itself more adequately. But before anything of the sort can be realized, the psychosocial possibilities of art and the best methods for realizing them in practice will need to be intensively studied.

So we come to man's second main higher activity—science. We must beware of the misuse of words like *science* and *scientific*, especially by those who want to cash in on the prestige of science to advance their own views or interests. Thus theology once arrogated to itself the title of Queen of the Sciences, and still claims rather plaintively to be a science—a claim which it could only justify by adopting scientific method. The most obvious modern example is that of the Marxists. Marx himself asserted that he had discovered the iron scientific laws that inevitably rule the development of society, and many people still accept Communism as a political creed because they have been told that it is "scientific". In actual fact Marxism is no more a science than theology—largely because it is itself a kind of theology, in the sense that it consists of a body of doctrines whose truth is guaranteed by dogmatic authority instead of being constantly tested against fact, and relies on a narrow and arrogant scholastic logic instead of on the patient humility of free enquiry.

Science, like art, is a loose and general term for a broad range of human activities and their products. Though its growing core is firm and clear, it is thus inevitably fluffy at the edges, and grades imperceptibly into non-science, as art does into non-art. It is perhaps best thought of as a process— the process of discovering, establishing, and organizing knowledge. To do this effectively, it must rely on scientific method.[1]

Looked at in the long perspective, science is seen as the continuation by new methods of the trend towards fuller and better-organized awareness which runs through the whole of animal evolution, from before the dawn of anything that could be called mind or memory, up to mammals and men. This trend was fostered by natural selection because it was biologically useful. Fuller and better-organized awareness enables its possessors to cope better and more fully with the changes and chances of their lives and their environment. In particular, it is a time-binder, enabling them to utilize past experience to guide future action.

Science has two interrelated psychosocial functions: it increases both comprehension and control. It enlarges man's understanding of the world, both the strange world of external nature and the equally strange world of his own internal nature; and it increases his capacity to control or guide various aspects and processes of those worlds.

As a result, everything in psychosocial evolution which can properly be called advance, or progress, or improvement, is due directly or indirectly to the increase or improvement of knowledge.[2]

Science is not merely a discovery of pre-existent facts: it is also, and more importantly, a creation of something new. It

[1] See e.g. J. Bronowski and B. Mazlish, *The Western Intellectual Tradition*, London, Hutchinson; New York, Harper, 1960, and C. G. Gillispie, *The Edge of Objectivity*, Princeton, 1960.

[2] The increase of knowledge is, of course, also responsible for much that is an obstacle to advance or even the reverse of it (a fact partly symbolized in the legend of the Fall). But this in no way impairs the validity of the fact that knowledge is the necessary basis for improvement.

99

is just as creative as art, though in a different way. Scientific laws are not something existing from eternity in their own right or in the mind of God, waiting to be discovered by man: they did not exist before men of science formulated them. The same is true of scientific concepts, like *atom*, or *electrical potential*, or *evolution*.

Scientific laws and concepts alike are organized creations of the human mind, by means of which the disorderly raw material of natural phenomena presented to crude experience is worked into orderly and manageable forms. A scientific concept is an intellectually effective integration of experience just as a painting is an aesthetically effective one.

Thus science is not only concerned with discovering facts: it is much more concerned with establishing relations between phenomena. Scientific comprehension was increased by relating the supposedly opposed qualities of heat and cold in the common concept of a scale of temperature; by bringing a number of apparently unrelated physical activities in relation with each other through the principle of the conservation of energy; by employing the concept of metabolism to perform the same service for a number of biological activities. A good scientific theory brings together a swarm of separate phenomena and their attendant concepts in a single unified pattern of relatedness. Modern evolution theory, for instance, has spun a comprehensive web of relations between the phenomena of cytology, genetics, adaptation, palaeontology, reproduction, embryology, behaviour, selection, systematics, and biochemistry.

Science is also concerned with understanding the systems of relatedness to be found in nature. This means the study of organization on every level—the level of atoms, of molecules, of individual organisms, of societies, of ecological communities. That being so, science cannot be only a matter of analysis, as if often erroneously supposed. It must start from the organizations to be found on any level. After studying them descriptively, comparatively and functionally, it can then try to analyse them into lower-level elements, and finally attempt a theoretical re-synthesis. It is no good trying to start from a lower level. Nobody could have built up the triumphant principles of modern genetics merely from a knowledge of biochemistry: genetic theory had to start with phenomena on the biological level, like mitosis and mendelian segregation, taking the facts of biological organization for granted. Only much later was it possible to analyse and understand genetic phenomena in biochemical terms, as we can now begin to do, thanks to Watson and Crick's brilliant theory of the self-replication of certain kinds of nucleic acid molecules.

We must beware of reductionism. It is hardly ever true that something is "nothing but" something else. Because we are descended from anthropoid primates, it does not follow that we are nothing but developed apes: because we are made of matter, it does not follow that we have nothing but material properties. An organization is always more than the mere sum of its elements, and must be studied as a unitary whole as well as analysed into its component parts.

Science is a self-correcting and self-enlarging system. It aims to unify experience. It creates patches of organized knowledge in the vast expanse of human ignorance. The patches of knowledge grow, and may fuse to form more comprehensive patterns. The trend is clearly towards an eventual single organization of conceptual thought, holding all aspects of experience in its web of relations, uniting all the separate patches of knowledge into one living and growing body of organized understanding. But meanwhile great gaps of ignorance still separate some of the partial systems, some of which are still isolated islands scientifically cut off from their neighbours, while some areas of experience are still recalcitrant to the method of science and remain outside its system.

The immediate need is for the scientific study of values. Philosophers and theologians sometimes assert that this is impossible, claiming that values lie outside the range of science. The Humanist cannot accept this: after all, values are phenomena, and therefore capable of being investigated by the methods of science. They are phenomena which only appear on the psychological level, and accordingly science must first approach them on this level. It must ask appro-

priate questions about them: In what psychological circumstances do values come into being? Out of what raw materials
are they constructed by man's psychometabolic activity?
What functions do they perform in psychosocial evolution?
How do they change and evolve? And just as science had to
devise special methods for dealing adequately with multi-
causal phenomena, especially where they are not amenable to
experiment, so, as time goes on, it will have to devise special
methods for dealing effectively with phenomena with a strong
subjective component. But a successful beginning has
already been made.

The study of values is a part of the one really major
problem now before science—the problem of relating mind
and mental activities to the rest of the phenomenal universe
in a single scientific picture. Here too there is much hard
work ahead; but here too a considerable measure of success
has already been achieved, partly through the evolutionary
study of animal behaviour, partly through the developmental study of human behaviour, partly through a joint
physiological and psychological attack on human mental
activity.

The irresistible trend towards the creation of one comprehensive scientific picture of the world of man's experience
emerges even more clearly when we look at science historically. Since the dawn of the scientific revolution some three
hundred and fifty years ago, science has steadily invaded new
fields. First of all, mechanics, astronomy, physics; then
chemistry and natural history; next followed geology and
physiology and embryology, and then experimental and
evolutionary biology; next was the turn of ethnology, then
psychology, then sociology. Science then proceeded to establish a footing in new territories like economics, archaeology,
and social anthropology, and established connections between
various separate disciplines with the aid of bridging sciences
like biochemistry, social psychology, epigenetics, and astro-
physics. We are now witnessing the invasion of the field of
psychosocial phenomena by science.

The only field still remaining outside the range of the
scientific system is that of so-called paranormal phenomena

like telepathy and extra-sensory perception (E.S.P.). If and when they are brought within its scope, some pretty radical alterations will presumably become necessary in its theoretical framework.

Meanwhile, however, science has attained a new and very real unity and firmness of organization and is giving us a scientifically-based picture of human destiny and human possibilities. For the first time in history, science can become the ally of religion instead of its rival or its enemy, for it can provide a "scientific" theology, a scientifically-ordered framework of belief, to whatever new religion emerges from the present ideological disorder.

This is imperative, since theology in this broad sense is a statement of beliefs and of their intellectual or rational justification: it dictates the general approach and character of a religion, as well as determining many of its particular features. Thus a theological system is to a religion what a framework of hypotheses and theories is to a science.

All theistic religions are based on the God hypothesis (or, to use Ralph Turner's more inclusive term, the daimonic hypothesis)—the belief that there exist supernatural beings of a personal or super-personal nature, capable of influencing natural events, including events in human minds. This is a dualistic theory, for it implies the existence of a basic and essential cleavage between natural and supernatural realms of being.

Early theologies are all polytheistic. Christian theology calls itself monotheistic, but permits itself a partial polytheism in the doctrine of the Trinity; while the position ascribed to the Virgin, the Angels and the Saints in Catholicism and to a lesser degree in other sects, gives full rein to polydaimonism. Christian theology bases itself on revelation and on belief in the historical reality of supernatural events such as the incarnation and resurrection of Jesus as the Son of God. It also maintains the reality of miracles.

A theological system incorporating such beliefs has a number of consequences which Humanists find undesirable. The belief in supernatural beings capable of affecting human destiny leads to petitionary rather than aspirational prayer, and to all kinds of propitiatory practices, from the use of incense to the bequeathing of rich gifts, from asceticism to penitential sacrifice. Belief in a supernatural after-life leads to concentration on attaining salvation in the other world and to a lack of concern for life in this world and its possible improvement. Belief in the fall of man and the necessity of redemption through an incarnate divine Saviour has led to the cruel (and untrue) doctrines of Original Sin and Damnation for unbelievers, as well as to a belief in the guilt and inherent inferiority of the female sex. Belief in the value of orthodox Christian beliefs and practices as the sole or main means of achieving salvation leads to the rejection or playing down of other ideas as to what constitutes "salvation", and of other methods of transcending selfhood. Belief in the Bible as the inspired word of God, and in the Church and its representatives as the sole source of correct doctrine, leads to a regrettable dogmatism and to the rejection or playing down of secular knowledge and scientific method.

Belief in a supernatural Ruler, endowed with absolute wisdom and the capacity of issuing absolute moral edicts, coupled with an ignorance of the workings of the unconscious as revealed by modern psychology, permits would-be dictators, fanatical moralists and other power-hungry men to believe that their subjective feelings of internal certainty are "really" the voice of an objective and external God and to claim divine guidance and sanction as a convenient disguise for their ambitions, and enables them with a good conscience to project their own guilt and resentful inferiority on to their enemies, and to canalize their repressed sadism on to their victims. How unfortunate for mankind that the Lord is reported by Holy Writ as having said "Vengeance is mine!"

Belief in the efficacy of ritual practices for ensuring salvation or other kinds of religious advancement has a deadening effect on the religious and moral life. Belief in supernaturalism and the miraculous and magical elements which go with it always leads to gross superstition, and usually to its financial exploitation. Think of the cult of relics, the complete repudiation of any scientific approach shown by the promulgation of doctrines like the bodily assumption of the Virgin Mary, by the proclamation of the miracle of Fatima, or by highly profitable pilgrimages to sites of "miraculous" cures like Lourdes.

Such theistically-based beliefs in various combinations can lead to a materialistic degradation of religion, sometimes silly, as in the prayer-wheels of Tibetan Buddhism; sometimes serious, as in the scandal of indulgences which started off the Reformation; and sometimes horrible, as in the human sacrifices of the Aztecs and the Carthaginians.

Above all, belief in an omnipotent, omniscient and omni-benevolent God leads to a frustrating dilemma at the very heart of our approach to reality. For many thinking people, it is incompatible with our knowledge of nature and history and with the facts of evil, suffering, and human misery. Even when, as in some modernist versions of Christian theology, the idea of a personal God is watered down and trans-mogrified into some abstract principle or supposed Absolute behind phenomena, and the Deity is removed farther and farther from any possibility of active interference in natural or human events, the dilemma remains. The human mind and spirit is not interested in such a Pickwickian God, and refuses to be fobbed off by assertions as to our inherent incapacity to understand Him. The theologian's assertion of divine incomprehensibility does not satisfy man in his modern world any more than Humpty Dumpty's remark, "Impenetrability, that's what I say," satisfied Alice in her Wonderland.

To sum up, any belief in supernatural creators, rulers, or influencers of natural or human process introduces an irreparable split into the universe, and prevents us from grasping its real unity. Any belief in Absolutes, whether the absolute validity of moral commandments, of authority of revelation, of inner certitude, or of divine inspiration, erects a formidable barrier against progress and the possibility of improvement, moral, rational, or religious. And the all-too-frequent combination of the two constitutes a grave brake on human advance, and, by obfuscating all the major problems of existence, prevents the attainment of a full and comprehensive vision of human destiny.

All this merely spells out the consequences of the fact that theistic religions, with their inescapable basis of divine revelations and dogmatic theologies, are today not merely incompatible with human progress and the advance of human knowledge but are obstacles to the emergence of new types of religion which could be compatible with our knowledge and capable of promoting our future progress.

Although destructive criticism of established religious systems, such as that of orthodox Christianity by militant Rationalism around the turn of the nineteenth century, may be necessary at certain periods, the time for negative activities is now past. It was not for nothing that Goethe made the Devil proclaim himself as *der Geist der stets verneint.*

What the world now needs is not merely a rationalist denial of the old but a religious affirmation of something new. However, it is harder to affirm, at least to affirm anything of lasting value, than to deny. It is harder for the same reason that, as the world has experienced on a gigantic scale, it is easier to destroy than to construct, easier to smash a cathedral, a city or a statue than to create one.

Construction needs a positive plan of some sort to work to and co-operative effort for its execution, and this demands intelligence, imagination, goodwill, and above all vision.

One of the main things needed by the world today is a new single religious system to replace the multiplicity of conflicting and incompatible religious systems that are now competing for the spirit of man. Our new vision of the universe and man's role in it is beginning to indicate the lines of its construction.

All religions, as I pointed out earlier, are psychosocial organs of evolving man: their function is to help him to cope with the problems of his destiny. They themselves evolve. But they always involve the emotion of sacred mystery experienced by men confronted with what Otto calls the numinous, the *mysterium tremendum*; the sense of right and wrong; and feelings of guilt, shame, or sin. They are always concerned in some way or another with the relation between the individual and the community, and with the possibility of his escaping from the prisoning immediacies of space, time, and selfhood by relating himself to some broader frame of reference, or in some self-transcending experience of union or communion with a larger reality.

They always possess what we may broadly call an ideology, a morality, and a ritual—an intellectual framework of beliefs, myths, and theological principles, an ethical framework of moral codes and injunctions, and an expressive framework of actions expressing or enhancing religious emotion.

As I set forth at greater length twenty-five years ago in my *Religion Without Revelation*, the raw materials out of which religions are formed consist of actual religious experiences, numinous or holy, mystical or transcendent. But the particular form which they take is primarily the result of their ideological framework of belief: I have given various examples of how the morality and the ritual expressions of a religion are determined by its beliefs to a much greater extent than its beliefs are determined by its morality or ritual.

Let us look at some of the major ideas which our new vision will contribute or dictate to the new belief-system. In the first place we have a totally different view of the mysterious. With the advance of scientific knowledge, many phenomena which once appeared wholly mysterious can now be described or explained in rationally intelligible or naturalistic terms. This applies not only to physical phenomena like rainbows and eclipses, pestilences and earthquakes, but also to biological phenomena like reproduction and sex, heredity and evolution, and to psychological phenomena such as obsession and possession, insanity and inspiration.

The clear light of science, we are often told, has abolished mystery, leaving only logic and reason. This is quite untrue. Science has removed the obscuring veil of mystery from many phenomena, much to the benefit of the human race: but it confronts us with a basic and universal mystery—the mystery of existence in general, and of the existence of mind in particular. Why does the world exist? Why is the world-stuff what it is? Why does it have mental or subjective aspects as well as material or objective ones? We do not know. All we can do is to admit the facts.

This means that, as Margaret Fuller said, we accept the universe. In spite of Carlyle's comment, "Gad, she'd better", this is not easy: there is great resistance to such acceptance. Initially, the universe reveals itself as too vast and varied to be accepted as a unitary whole by our small human minds; many of its components are apparently incommensurable with human thought and feeling, and in many of its aspects it appears alien and even hostile to human aspiration and endeavour. But we must learn to accept it, and to accept its and our existence as the one basic mystery.

Accordingly, any new emergent religion must have a background of reverence and awe in its belief-system, and must seek to keep alive man's sense of wonder, strangeness and challenge in all his particular dealings with the general problem of existence.

But though all we can do about the universe in its total existence is to discover it as an irreducible mystery, to be humanly assimilated only by wonder and free acceptance, yet the details of its phenomenal working and the relations of its operative parts can be profitably clarified by human intellectual and imaginative effort. And this applies to religion as well as to science or to art. In all of them the ecological approach is essential.

Religion can be usefully regarded as applied spiritual ecology. The relations with which a religion must attempt to deal are the relations of mankind with the rest of external nature, the relation of man's individualized self with the rest of his internal nature, and the relation of individual men and women with other men and women and with their communities.

All these can be much clarified by our new humanist vision. In its light the universe is seen as a unitary and evolutionary process. Man is part and a product of the process, but a very peculiar part, capable of affecting its further course on earth and perhaps elsewhere. But he is only able to affect the process constructively by understanding its workings.

The rightness of relation he must aim at has two aspects. One is a relation of right position in an integrated and harmonious pattern; the other (and this is the major novelty introduced by the new vision) is a relation of right direction with the whole process. Man's religious aim must therefore be to achieve not a static but a dynamic spiritual equilibrium. And his emergent religion must therefore learn how to be an open and self-correcting system, like that of his science.

All religions provide for some ceremonial sanctification of life, especially of events like birth, marriage and death, and those marking the transition from one stage of life to another, like initiation or the taking of a degree: his new emergent religion must continue to do this, though it must translate the ceremonials into terms that are relevant to the new vision and the new circumstances of his life.

This reformulation of traditional religious concepts and beliefs and ceremonies, their translation into a new terminology and a new framework of ideas, is a major task for Humanism.

Man makes his concepts. He constructs them out of the raw material of his experience, immediate and accumulated, with the aid of his psychological machinery of reason and imagination. This is true not only of religious concepts but of scientific concepts like the atom or natural selection today, or the four elements or the inheritance of acquired characters in earlier times. But whereas science is constantly and willingly improving its terminology and reformulating its concepts, even scrapping them and constructing quite new ones, religion on the whole resists any such transformation.

Religious concepts like god, incarnation, the soul, salvation, original sin, grace, atonement, all have a basis in man's experiences of phenomenal reality. It is necessary now to analyse that basis of reality into its component parts, and then to reassemble these elements, together with any new factors that have come to light, into concepts which correspond more closely to reality and are more relevant to present circumstances.

Thus, if I may over-simplify the matter, *God* appears to be a semantic symbol denoting what Matthew Arnold called "the power not ourselves", or rather the various powers felt to be greater than our narrow selves, whether the forces of external nature or the forces immanent in our own nature, all bound together in the concept of a personal or super-personal sacred being in some way capable of affecting or guiding or interfering in the course of events. The forces are real enough: what we have done is, quite illegitimately, to project the god concept into them. And in so doing we have distorted their true significance, and effectively altered the course of history.

Once this is realized, it should be possible to reformulate such ideas as Divine Law, obedience to God's will, or union with the mind of God, in an evolutionary terminology consonant with existing scientific knowledge.

Again, Christian ethics (to which the world owes a great debt) are based on the doctrine of Original Sin resulting from the Fall of Man. This is an attempt to provide an acceptable interpretation of such general and wellnigh universal phenomena as our sense of guilt, our search for atonement and for some form of salvation, our authoritarian consciences, our rigorous sense of right and wrong, our consequent persecution of those who deviate from what we feel is the right path.

As I attempted to show twenty years ago in my Romanes lecture on *Evolutionary Ethics,* and as Professor Waddington has successfully demonstrated with a wealth of supporting argument in his recent admirable book, *The Ethical Animal,* psychology and evolutionary biology between them are now indicating a rational and coherent explanation for these facts.

Psychosocial life is based on the transmission of accumulated experience in the form of tradition. And this, Waddington makes clear, cannot be effective unless the human infant is genetically equipped as an "authority-acceptor": he is constructed so as to accept what he is told by his parents as authoritative, in the same sort of way as baby birds are equipped with an imprinting mechanism which makes them accept any moving object within certain limits of size as a parent.

This "proto-ethical mechanism" involves the internalization of external authority in the baby's primitive conscience, a process accompanied by all-or-nothing repression of impulses of hate for the authority who is also the loved parent. As a result, a quality of absoluteness becomes attached to the baby's sense of rightness and wrongness, together with an ambivalent attitude to authority in general: his morality is burdened with a load of guilt, and his feelings towards authority become impregnated with ambivalence.

All this happens before he is old enough to verify his ideas by experience. During his later development he will modify and rectify the content and authoritarianism of what he has accepted, but will generally retain a great deal of both. The aim of the Humanist must be, not to destroy the inner authority of conscience, but to help the growing individual to escape from the shackles of an imposed authority-system into the supporting arms of one freely and consciously built up. And this will involve a thorough reformulation of the ethical aspects of religion.

Reformulation—even reappraisal—is perhaps most necessary in regard to man's inner life and what, for want of a better terminology, is called spiritual development.

Religious experiences such as those of communion with some higher reality, or inspiration from outside the personality, or a sense of transcendent power or glory, or sudden conversion, or apparently supernormal beauty or ineffable sacredness, or the healing power of prayer or repentant adoration, or, above all, the deep sense of inner peace and assurance in spite of disorder and suffering, can no longer be interpreted in the traditional terms of communication with a personal God or with a supernatural realm of being. But neither can they be denied or explained away by over-zealous rationalism as merely illusory products. They are the outcome of human minds in their strange commerce with outer reality and in the still stranger and often unconscious internal struggle between their components. But they are none the less real and they can be of great importance to the individual who experiences them: but further, as the Churches well know, they need to be examined and disciplined.[1]

[1] Besides William James's famous book, there are many valuable descriptions and studies of the varieties of religious experience, a number of which I have cited and discussed in my *Religion Without Revelation*.

Religious experiences often are or appear to be ineffable in the literal sense of the word, which makes their discussion very difficult. But their significance is a matter both high and deep (as I am in all humility aware); and they certainly need re-examination and reappraisal if their great potential value is to be realized.

Further, experiential religion should enlist the aid of psychological science in a radical study of man's actual and potential spiritual resources. Such a study would, of course, have to start from the presuppositions that "man" is a new type of organism consisting of individual mind-bodies interacting with a superindividual and continuing system of ideas and beliefs, whose destiny is to actualize more and more of his possibilities for greater fulfilment during further evolution; and that "religion" is an organ of man primarily concerned with what is felt and believed to be sacred or transcendently important in that destiny.

But our new vision illuminates our existence and our destiny in a new way, and necessitates a new approach to their problems. In its light we see at once that the reappraisal of religious experience must be a part of something much larger—a thorough investigation of man's inner world, a great project of "Mind Exploration" which could and should rival and surpass "Space Exploration" in interest and importance. This would open up a new realm of being for colonization and fruitful occupation by man, a realm of mental realities, built on but transcending the realm of material realities, a world of satisfactions transcending physical satisfactions, in some way more absolute and more perfect. Ordinary men and women obtain occasional glimpses of it through falling in love, or through overwhelming experiences of ecstasy, beauty or awe. And we have the reports of the occasional mental explorers, poets, thinkers, scientists and mystics who have penetrated into its interior. Think of St Teresa, or of Blake as the Mental Traveller, or of Wordsworth anticipating Freud by revealing in us the "high instincts before which our mortal nature Doth tremble like a guilty thief surprised".

No concerted effort has yet been made towards its

exploration or adequate mapping. There is as yet no proper terminology for its discussion. In describing its workings and results, ordinary language falls back on terms like *rapture* and *inspiration, magical* and *heavenly, bewitching* and *divine*, while the first attempts at scientific terminology, like *repression* and *sublimation, id* and *superego*, deal only with its fringes.

From the specifically religious point of view, the desirable direction of evolution might be defined as the divinization of existence—but for this to have operative significance, we must frame a new definition of "the divine", free from all connotations of external supernatural beings.

Religion today is imprisoned in a theistic frame of ideas, compelled to operate in the unrealities of a dualistic world. In the unitary Humanist frame it acquires a new look and a new freedom. With the aid of our new vision, it has the opportunity of escaping from the theistic impasse, and of playing its proper role in the real world of unitary existence.

This brings me back to where I started—to our new and revolutionary vision of reality. Like all true visions it is prophetic; by enabling us to understand the present condition of life in terms of its extraordinary past, it helps us not only to envisage an equally extraordinary future but to inject planned purpose into its course.

In its light, fulfilment and enrichment of life are seen as the overriding aims of existence, to be achieved by the realization of life's inherent possibilities. Thus the development of man's vast potential of realizable possibility provides the prime motive for collective action—the only motive on which all men or nations could agree, the only basis for transcending conflicting ideologies. It makes it possible to heal the splits between religion and science and art by enlisting man's religious and scientific and artistic capacities in a new common enterprise. It prescribes an agenda for the world's discussions of that enterprise and suggests the practical methods to be employed in running it.

It indicates the urgent need for survey and research in all fields of human development. This includes the promotion of what I may call a psychosocial technology, including the production of ideological machine-tools like concepts and beliefs for the better processing of experience.

We also need to develop a new ecology, an ecology of the human evolutionary enterprise. This means thinking out a new pattern of our relations with each other and with the rest of our environment, including the mental environment which we both create and inhabit.

Psychosocial ecology must aim at a right balance between different values, between continuity and change, and between the evolutionary process for whose guidance we have responsibility and the resources with which we have to operate. Those resources are of two kinds—material and quantitative, for maintenance and utility; and psychological and qualitative, for enjoyment and fulfilment—such things as food, and energy, mines and industrial plants on the one hand; solitude, landscape beauty, marine and mountain adventure, the arts, the wonder of wild life on the other. Planned human ecology must balance and where possible reconcile our claims on the two kinds of resource.

What is the place of the individual in all this? At first sight the individual human being appears as a little, temporary, and insignificant creature, of no account in the vast enterprise of mankind as a whole. But in Evolutionary Humanism, unlike some other ideologies, the human individual has high significance. Quite apart from the practical function which he performs in society and its collective enterprises, he can help in fulfilling human destiny by the fuller realization of his own personal possibilities. A strong and rich personality is the individual's unique and wonderful contribution to the psychosocial process.

*　　*　　*

Santayana has come close to the central idea of Evolutionary Humanism in sane and splendid words. "There is only one world, the natural world, and only one truth about it; but this world has a spiritual life in it, which looks not to another world but to the beauty and perfection that this world suggests, approaches and misses."

If we aspire to realize this potential beauty and perfection

114

more fully, we shall have to utilize all the resources available —not only those of the external world, but those internal resources of our own nature—wonder and intelligence, creative freedom and love, imagination and belief. The central belief of Evolutionary Humanism is that existence can be improved, that vast untapped possibilities can be increasingly realized, that greater fulfilment can replace frustration. This belief is now firmly grounded in knowledge: it could become in turn the firm ground for action.

But it is time to bring this lengthy argument to a summary conclusion. Increase of knowledge leads to new idea-systems—new organizations of thought, feeling and beliefs. Idea-systems in this sense provide the supporting framework of human societies and cultures and in large measure determine their policies and course. During human history (psychosocial evolution), the adoption of each new type of idea-system has initiated a new type of society, a new step in psychosocial evolution.

At the moment, the increase of knowledge is driving us towards the radically new type of idea-system which I have called Evolutionary Humanism. The position is critical, because the guidance of this new type of idea-system is needed to prevent psychosocial evolution from becoming self-defeating or even self-destroying.

The immediate effort needed is an intellectual and imaginative one—to understand this new revelation made to us by the growth of knowledge. Humanism is seminal. We must learn what it means, then disseminate Humanist ideas, and finally inject them whenever possible into practical affairs as a guiding framework for policy and action.

EDUCATION AND HUMANISM

W H A T is education? Dorothy Parker once said that it was casting sham pearls before real swine. But this splenetic outburst is hardly fair, and could only represent the views of a harassed and overburdened teacher in a difficult neighbourhood. My grandfather, Thomas Henry Huxley, once defined an educated man in the following famous passage:

> That man, I think, has had a liberal education who has been so trained in youth that his body is the ready servant of his will, and does with ease and pleasure all the work that, as a mechanism, it is capable of; whose intellect is a clear, cold, logic engine, with all its parts of equal strength, and in smooth working order; ready, like a steam engine, to be turned to any kind of work, and spin the gossamers as well as forge the anchors of the mind; whose mind is stored with a knowledge of the great and fundamental truths of Nature and of the laws of her operations; one who, no stunted ascetic, is full of life and fire, but whose passions are trained to come to heel by a vigorous will, the servant of a tender conscience; who has learned to love all beauty, whether of Nature or of art, to hate all vileness, and to respect others as himself.

Elsewhere, in the same address, after enumerating the biassed arguments of the politicians, the clergy, the manufacturers and the capitalists, he wrote with deep feeling of education's central mission of enlightenment: "A few voices are lifted up in favour of the doctrine that the masses should be educated because they are men and women with unlimited capacities of being, doing, and suffering, and that it is as true now, as ever it was, that the people perish for lack of knowledge."

But I want to treat education as a social process. For me, education is an organ of man in society, whose basic function is to ensure the continuity and further advance of the evolutionary process on earth by the transmission and transformation of tradition. It exists in rudimentary form in higher vertebrates, but only becomes of major importance in our own species, where it transmits the knowledge, the skills and beliefs, the attitudes and ideas which are necessary for the maintenance, achievement and development of man in society. Much of education in this broad sense is unorganized, acquired through press and radio and public meetings, or through individual self-education. Here I shall confine myself to organized systems of education (though clearly these can be so planned as to encourage and gear in with various kinds of unorganized education).

To look at education in the perspective of evolution is good for many reasons. Firstly, it shakes us out of over-preoccupation with the multitude of specific and immediate difficulties that beset us. Secondly, it provides a necessary corrective to the tendency of our neotechnic industrial civilization to think and plan in terms of quantity rather than quality—a tendency which is having unfortunate effects on education, both on subject-matter (think of the emphasis on the physico-mathematical sciences and technology as against the life and earth sciences and conservation), on curriculum (think of the emphasis on subjects which can easily be examined in), on evaluation (think of the emphasis on examination standards and marking systems) and on methods (think of teaching aids). Finally, while stripping education of much sentimental and idealistic mystique, it gives new dignity and importance to the whole educational process, new inspiration for all those engaged in teaching, new clarity to the aims and principles of education.

Education involves the transmission of experience and its results across the gap from earlier to later generations. There are three stages of its evolution, two of them prehuman. In the most primitive, transmission bridges only one generation-gap between parents and offspring. This stage is found most obviously in carnivores like fox or lion, where the young animal, after a period of self-education by play, learns to hunt by accompanying one or both of its parents.

The second stage is marked by the beginnings of tradition, a set of social habits peculiar to a single population, which endures through a number of generations. Rudiments of such proto-tradition are found in various social mammals (and perhaps a few birds); but the most striking example is that of the Japanese monkey, *Macaca fuscata*, summarized by Dobzhansky in his book, *Mankind Evolving*: in this species, each monkey colony or social group has its own traditional food-habits, but these are occasionally modified by adventurous innovations, which take two or three years to spread through the group. Here we see how even a proto-tradition can evolve by incorporating new elements. But it involves no real educational system—no special arrangements for educating the young animals in and through the tradition.

The third stage is found only in man. It is the stage of a truly cumulative tradition, based on conceptual thought and communicated by artificial or arbitrary symbols that have to be learnt, as well as by imitation and by simple signs that are genetically determined. As the tradition grows more complex and its symbolic system of communication correspondingly more elaborate, organized education becomes necessary as part of the machinery for its transmission down the generations.

Cumulative tradition has enabled man to cross the critical threshold from the biological to the psychosocial phase of the universal evolutionary process. In this new phase, evolution is manifested primarily in cultural change, only secondarily in change in genetic constitution. To take one example, only negligible genetic changes in man's intelligence or other capacities have taken place since the Neolithic period (and quite possibly since Cro-Magnon times), but almost incredible changes in his cultural apparatus and achievements, such as religion, art, science, law, technology, social organization, literature and education.

One of the novel characteristics of the psychosocial phase is its new tempo of change. Not only is cultural evolution much faster than biological, but it shows acceleration. In the Lower Paleolithic it took perhaps 100,000 years to effect a major cultural change; in early civilizations a few hundred years; and now barely a single decade—bringing new problems for education every few years.

And yet, in spite of this new speed of change and of the astonishing new achievements that it has permitted, we must remember that man is an extremely recent phenomenon. Man as a hominid in the broad sense, including the ape-man Pithecanthropus and his immediate ancestors, cannot be much more than a million years old, and a million years in the perspective of life's two and three-quarter thousand million years is less than half a minute out of a twenty-four-hour day; while the whole of civilized man's history has all been compressed within one tick of the cosmic clock. What is more, he is an unfinished type, needing a great deal of improvement before he can qualify as a satisfactory product of evolution. Human individuals are all too obviously unfinished when they are born into the world: they need radical improvement before they can qualify as satisfactory social beings. And their improvement as individuals, as well as the improvement of man as a type, depends to an important extent on the improvement of education.

We are only beginning the detailed scientific exploration of the basic agencies and mechanisms of human evolution, but some broad principles are emerging. Those most relevant to my subject are these. First, that it is the general idea-system, the ideological pattern of ideas and knowledge, values and beliefs, which essentially characterizes human societies and cultures, and indeed is the prime determiner of their social organization, their material substructure and their achievements. The psychosocial process is a unitary one, simultaneously and indissolubly both material and mental, in which the material machinery and the social structure are constantly reacting back on the ideological pattern. But mind dominates the process, and it is the ideological pattern which is primary.

Secondly, with the extension of our understanding of the psychosocial process now and in the past it is becoming clear that everything in human history which deserves the title of advance or progress, every improvement or hope of improvement of the human lot and man's hard destiny, has sprung from the discovery and dissemination of new knowledge and new ideas.

Let me clarify this point a little I am using *ideas* in a broad and loose sense to cover beliefs and values as well as general and scientific concepts. Ideas are real phenomena, part of man's psychosocial equipment. Before new factual knowledge can become effective it must be organized, incorporated in, or related to an idea. Ideas, of course, grow out of our experiences, both external and internal, but differ a great deal in what may be called their scientific or objective validation. Where detailed and scientific validation is meagre or not available, ideas can still exact strong psychosocial effects: for instance, the idea of monotheism, or of personal salvation through religious faith or sacred ritual. But in the long run those ideas will prevail which are scientifically validated and are backed by a mass of established factual knowledge.

The psychosocial process is a cybernetic or self-directing one, full of feedback mechanisms, so that new ideas, and the new knowledge of which they are the vehicles, not only must operate within the contemporary idea-system, but will modify it. Thus the basically new idea of domesticating food-supply by growing crops interacted with the general ideas of the time and became part of a general ideological framework concerned with agricultural prosperity, fertility ritual and sacred priest-kingship, which in turn modified social structure.

*   *   *

The knowledge explosion of the last hundred years since Darwin is giving us a new vision of our human destiny—of the world, of man, and of man's place and role in the world. It is an evolutionary and comprehensive vision, showing us all reality as a self-transforming process. It is a monistic vision, showing us all reality as a unitary and continuous process, with no dualistic split between soul and body, between matter and mind, between life and not-life, no cleavage between natural and supernatural; it reveals that all phenomena, from worms to women, from radiation to religion, are natural.

It will inevitably lead to a new general organization of

thought and belief, and to the development, after centuries of ideological fragmentation, of a new and comprehensive idea-system. The Middle Ages had a comprehensive vision and a comprehensive idea-system, and so does Marxist Communism today; but neither was founded on comprehensive knowledge. The present is the first period in history when man has begun to have a comprehensive knowledge of stars and atoms, of chemical molecules and geological strata, of plants and animals, of physiology and psychology, of human origins and human history. The knowledge is highly incomplete; new and surprising discoveries are being made every year and will continue to be made for centuries to come. But it is comprehensive, in the sense of covering every aspect of reality, the whole field of human experience.

Its upshot is clear. Man is not merely the latest dominant type produced by evolution, but its sole active agent on earth. His destiny is to be responsible for the whole future of the evolutionary process on this planet. Whatever he does, he will affect that process. His duty is to try to understand it and the mechanisms of its working, and at the same time direct and steer it in the right direction and along the best possible course.

This is the gist and core of Evolutionary Humanism, the new organization of ideas and potential action now emerging from the present revolution of thought, and destined, I prophesy with confidence, to become the dominant idea-system of the next phase of psychosocial evolution.

What are the implications of this new pattern of thought and belief? Its overriding aim so far as I can see, must be defined positively, in terms of fulfilment and achievement—greater fulfilment (and therefore less frustration and misery) for more human individuals, interlocking with fuller achievement (and therefore less muddle and failure) by more human societies. And this can only be secured by a better understanding and a fuller realization of human possibilities. Further, in pursuing this aim, man must remember that he is a part of nature, and must learn to live in harmonious symbiosis with the environment provided by his planet, a relation of responsible partnership instead of irresponsible exploitation. If he is to make a success of his job as guiding agent for evolution, he must abandon the arrogant idea of conquering and exploiting nature; he must co-operate and conserve. Here, too, he must set himself to understand possibilities and try to realize them more fully and fruitfully, but here they are the possibilities not of human, but of external nature.

If this view of human destiny is essentially correct, then clearly our educational system and methods must come to terms with it, for after all education must be concerned with man's place and role in nature, and its raw material is man himself. As I wrote in my Huxley Lecture twelve years ago: "The most important, if not the most urgent, task of our times is the development of a new set of integrative, directive and transmissive mechanisms for human societies and for their continuity down the generations. These must include systems in which the community at large can share—systems of shared interpretation, shared belief, shared activity and shared faith."

Most educational systems are highly resistant to change, because they are controlled by dogmatic religious organizations, or because they are closely linked with the established social order, or just because of inherent conservatism. Today we need a radical change of system; and clearly the new system must itself be evolutionary, not change-resistant but change-promoting. It must transform as well as transmit. In part, this can be achieved through appeals to morality—by showing growing boys and girls a comprehensible but high aim in life, making them understand the moral duty of helping and guiding the evolutionary process in a desirable direction. But preaching is not enough: something more practical is needed. If, as I maintain, our essential aim be greater fulfilment, then the next step in psychosocial evolution must assuredly be from the Welfare State towards a Fulfilment Society. A humanist educational system will not only put the idea of the fulfilment society before children, but will provide them with opportunities for actual personal fulfilment in every possible way—through knowledge (I remember Bertrand Russell once exclaiming, "How nice it is to *know* things!"); through disciplined adventure on mountain or sea; through expeditions and travel; through painting and acting and making music; through enjoyment of nature and beauty; through fun and games; through inner peace; through study projects and organized discussions; through responsible participation in group activities. It will have a curriculum of experience as well as a curriculum of subjects.

In adapting our old educational system to our new vision, a lot of cargo will have to be jettisoned—once noble but now mouldering myths, shiny but useless aphorisms, Utopian but unfounded speculations, nasty projections of our prejudices and repressions. Thus, man was not created in his present form a few thousand years ago. Mankind is not descended from Adam and Eve, or any other single couple. Children are not born with a load of original sin derived from the Fall, nor with a *tabula rasa* of a mind ready to be inscribed with whatever message educators wish. There never was a Golden Age, nor a Noble Savage. There are no pure races, nor any Superior or Master Races. Mind and Body are not separate entities. There are no Absolutes of truth or virtue, only possibilities of greater knowledge and fuller perfection.

On the contrary, mankind is a single species, which originated from a population of ancestral apes about a million years ago. His evolution since then has been marked by the increasing complexity and improvement of his material, social and psychological organization, but at the same time by the increasing magnitude of his crimes and follies. Children are not a set of uniform *tabulae rasae* but highly complex and varied psychosocial organisms engaged on the extremely difficult task of developing into satisfied and satisfactory members of a social community.

The educational process has to cope simultaneously with several distinct problems—the imparting of knowledge, practical and theoretical; the learning of skills and social habits; the transmission of traditions and beliefs, religious and secular; the formation of character and personality; moral as well as intellectual development; and the crucial passage from childhood to responsible adult life.

123

In tribal societies, there is nothing that can be called a curriculum. Children gain the knowledge they need for the practical activities of life by watching and imitating the adults and by actual participation. Punishment is rare. What may be called their theoretical instruction, in tribal myth and morality and in forms of adult behaviour, is imparted by selected elders, often with the aid of esoteric and awe-inspiring ceremonies. This takes place during a special period of preparation, culminating in a painful and frightening ordeal which usually includes sexual mutilation. There is no examination to test their knowledge, only an ordeal to prove their adulthood; no certificate to frame and cherish, only the fact of acceptance as full members of the tribe.

Things have changed since then. But we still inflict a painful and frightening ordeal on our children, in the shape of a whole series of examinations; and we still celebrate their successful survival of these ordeals by ceremonial activities, most elaborately by the solemn pomp of university commencements and degree days. Most psychiatrists, I am sure, would agree that one important reason for the stiffness of examinations is psychological—the subconscious desire of the adult to revenge himself for past ordeals by subjecting the young to the same unpleasant trials to which he himself was subjected: and assuredly the gorgeous robes of the academic procession are worn not only to impress the community gathered for the occasion but as a permitted outlet for donnish egos (it certainly was so for mine!). And yet it is still true, as Professor Elvin has pointed out in his chapter on "An Education for Humanity" in *The Humanist Frame*, that a tribal boy's education may be a better preparation for life in a tribal society than is our education for life in our vaunted technological society.

There is need today for drastic change. Education must come to terms with Humanism over its curriculum, over its relations with society, and over its methods of fostering the development of personality.

As regards curriculum, the solution is, in principle, simple. The advance of knowledge has at last given us a unified or unitary picture of phenomenal reality: clearly the curriculum must reflect this unity and itself become unified instead of a number of unrelated fragments called *subjects* or *activities*.

Simple in principle, but not so simple in practice! The first difficulty is a quantitative one—that the mere growth of knowledge, not only scientific but also historical and sociological, is too large for any single curriculum to unify. However, the prime function of education is not to impart the maximum amount of factual information, but to provide comprehension, to help growing human beings to a better understanding of the world and themselves. And for this we need what I have broadly called *ideas*.

Ideas in this broad sense are mental machinery for dealing with bodies of experiential fact, the necessary tools of comprehension. Readers of Helen Keller's autobiography will remember the dramatic moment when the deaf, dumb and blind child suddenly realized that "everything has a name". This is the basic idea underlying human language and human thought; and for her it was the master key which unlocked her latent understanding. Ideas can facilitate understanding in boys and girls during their education just as much as in scientists in their research or professional men in their careers. But they must have a sufficiency of facts to work with: in education, the child must be provided with the right mixture of facts and ideas.

As with material machines, some ideas are better than others—they can handle more apparently disparate phenomena, can bridge larger gaps, can simplify greater diversity. Thus the single idea that the planets move round the sun in ellipses relegated all the fantastic ptolemaic machinery of cycles and epicycles to the scientific dustbin, and with one stroke simplified for all later generations the whole working of the solar system. Again, the abandonment of the scholastic idea of two conflicting principles of heat and cold in favour of the scientific idea of a single scale of temperature, was (and still is) a great aid to our understanding of physical phenomena.

Evolution—or, to spell it out, the idea of evolutionary process—is the most powerful and the most comprehensive idea that has ever arisen on earth. It helps us to understand our origins, our own nature, and our relations with the rest of nature. It shows us the major trends of evolution in the past and indicates a direction for our evolutionary course in the future.

Since the process of evolution involves the constant inter-adjustment of organisms with their environment, the evolutionary idea gives us a new view of our relations with our planetary habitat and its resources, and with the other organized life-communities with which we share it. It gives us a new perspective in time, and a new sense of universal interrelatedness. Above all, it unifies our knowledge and our thought. We are part of a total process, made of the same matter and operating by the same energy as the rest of the cosmos, maintaining and reproducing ourselves by the same type of mechanism as the rest of life, unique only in having been pushed further along life's general road to reach the psychosocial stage.

Thus the evolutionary idea must provide the main unifying approach for a humanist educational system, and evolutionary biology could and should become a central or key subject in its curriculum. As I said in an address to American high school teachers at the Darwin Centennial Celebration at Chicago in 1959: "Not only is evolution the necessary background for any proper understanding or exposition of biology; but I, with many of my colleagues, feel strongly that biology is the necessary basis for understanding ourselves and nature and our place in nature.

"Evolution is important for understanding ourselves as animal organisms, for instance in connection with food, health, and disease. Evolution is essential for understanding ourselves in relation to our environment and other organisms in that environment—in other words, for understanding human ecology. Evolution is also essential for understanding ourselves as organisms which develop—in other words, for understanding human embryology and ontogeny; the most spectacular phenomenon in life is the development of adult human beings from microscopic bits of nucleated protoplasm. Embryology links up with an understanding of human reproduction and with an understanding of that rather difficult but important subject, the genetic basis of our life. Finally, evolution helps us to understand ourselves as unique organisms equipped with a new *method of evolution*—cultural evolution—based on the cumulative transmission of experience through language and symbols.

"Embryology, reproduction, and genetics reveal all sorts of extremely exciting facts. 'Exciting' is the right word, for knowledge of these facts does excite interest and wonder in our minds. I use the word 'wonder' deliberately, for I believe that to excite wonder and interest in the variety and richness of life is important in education. So far as I can see, biology is the best scientific subject for eliciting a sense of wonder and an immediate interest in the strange, the unusual, and the exciting. Biology may not stir the interest of all the mechanically minded, but it does arouse the interest of a great many children, probably the majority.

"Biology has the further advantage that through it you can enlist the born naturalist as well as the born laboratory experimenter—the boy or girl who is interested in the variety and the wonder of things as they are, as well as the child who is interested in finding out how they work. It is no coincidence that all the great evolutionists have had an interest in natural history and that most of them started as naturalists."

Too often, perhaps especially in the U.S.A., owing to religious pressure or plain conformist timidity, evolution is not allowed into the curriculum, or is admitted under some specious alias, like "racial development"; whereas in the U.S.S.R., in spite of the setback due to Lysenkoism, evolution occupies an important place in education. But for further facts and ideas on this, I must refer my readers to Professor H. J. Muller's admirable article, "One Hundred Years without Darwinism are Enough".

Evolutionary biology provides us with another unifying idea of the greatest importance—the ecological idea. Ecology is scientific natural history; it is the science of relations *par excellence*—relations between organisms and their environment, and of organisms with each other. It helps us to understand how life makes a living.

It deals with what is commonly called the balance of nature. But its central concept is that of the organized ecological community—a patterned assemblage of different vegetational, animal and microbial types, inter-adjusted to give optimum utilisation of the resources of a particular habitat. If the dynamic pattern of relations is interfered with, the entire habitat and the ecological community living in and on it may be damaged. The most obvious case is deforestation with consequent erosion; another is the introduction of domestic cattle coupled with reduction of wild herbivores, which may disrupt the ecological pattern of savannah and turn it into desert. From earliest times, man has been interfering with more and more natural habitats and has exploited their resources more and more pitilessly. He has already converted vast areas of once fertile land, in the Mediterranean and the Middle East, in China and in India, into arid semi-deserts or treeless infertility. Now that his exploding population is subjecting more areas to ever more drastic and more technologically efficient exploitation, he is in danger of becoming the cancer of the planet instead of its guide.

In the educational systems of underdeveloped territories, children should be introduced to science by the biological way of ecology and physiology and their applications in conservation and health, not by way of physics and chemistry and their applications in technology and industry. And in all countries, ecology is essential as a basis for good land use and productive development.

Man lives in a triple tier of environments, material, social and psychological. Ecology in the customary sense deals with man's relations with the forces and resources of external nature. Social ecology deals with man's social relations, both within and between human societies. And what we may call psychological ecology is concerned with man's individual and collective relations with the forces and resources of his inner nature and the environment of ideas, beliefs and values which he has created and with which he has surrounded himself.

How should the new humanism's evolutionary approach take effect in education? The overriding need is that it

128

should put an end to the fragmentation of the present system. Education must be comprehensive, in dealing with every aspect of life; it must also have a unitary pattern, reflecting the unity of knowledge and the wholeness of experience. It must attempt to give growing minds a coherent picture of nature and man's role in it, and to help immature personalities towards integration and self-realization.

To give a coherent picture, we need in the first place an integrated curriculum instead of a patchwork, a curriculum focused on the pattern of man's relations with nature and the psychosocial process, instead of on separate aspects of nature like physics or botany, and on separate aspects and activities of man like literature or history.

Once we begin to think along these lines, we find that different "subjects" can link up with and reinforce each other. Physiology links up with chemistry and heredity, ecology with geography, soil science with agriculture and meteorology. The study of development leads on to biological evolution; evolution in turn links back to geology, astronomy and cosmology, and on to man's origins, to archaeology and human history. History in turn links up with economics, social studies and citizenship. Art and architecture, law and morality, science and technology are best treated as functions of man in society: they too evolve, and accordingly profit by being treated historically, not merely as separate and complete packages of facts and principles.

Just how to plan and introduce such a curriculum is a matter for the educational profession. Of course it will not be easy, in face of the competitive claims of traditional subjects and established university departments on the one hand, and on the other hand of the portentous growth of factual knowledge and theoretical scaffolding, notably in the sciences, but also in the humanities. It will mean scrapping a great deal of dead wood and dead weight: it will mean some sacrifice of specialists' *amour propre* and curricular claims: and it will take much ingenuity and a great deal of goodwill. But it must be done, and I am sure that it can be done.

Indeed, the process is already beginning. Medicine is throwing overboard a mass of unnecessary anatomical detail:

universities are introducing first-year courses of general studies: sixth forms are taking steps to correct earlier over-specialization: the American Institute of Biological Sciences has worked out a remarkable integrated curriculum for High School biology, embodying many new ideas and methods, and the Gulbenkian Foundation is doing something similar in this country. But these are sporadic and isolated attempts. The task will only be satisfactorily achieved when we realize that, here as elsewhere, unity brings strength. In a properly unified curriculum, separate subjects will not compete for prestige and place, but can reinforce each other.

This could give a new dynamic to the educational process. Children in general have a natural interest and curiosity: they want to know more about this strange and wonderful world, about human life and how to live it, and to find out what it all means. If their education is designed to help them in this instead of setting up a number of hurdles in their way, they will enjoy it, and will become willing co-operators instead of reluctant or obstructive victims. And education will acquire a new social dynamic: from being merely a pre-paration for life, it will become an integral part of life, an instrument of man's evolution.

The other prerequisite for a humanist education is that it should help children to realize more of their capacities, and to develop into well-integrated personalities. For this, in addition to the integration of subjects, we need an integration of activities. The educational system must set itself not merely to surmount obstacles—to bypass frustration, to over-leap apathy, to resolve conflict—but to provide opportun-ities, for active living, for satisfying achievement, for feeling significant, for fulfilment. This again will not be easy. Apart from difficulties of staff and facilities, and the difficulty of linking school life, home life and social life, there is the difficulty of time. Once more, however, it not only must, but can be done.

Here again, unification leads to mutual reinforcement. So-called extra-curricular activities can be made to serve curricular ends, and vice versa. Children in general have a natural desire to achieve, to exercise their capacities, to count

for something in life. If they find that their education is designed to help them in this instead of treating them as temporary
inmates, to be dealt with by a combination of bribery and
punishment, they will become willing participants.

A further important implication of the evolutionary-
humanist approach is that the mind-bodies the educator has
to deal with are not all alike. Far from it! They are extremely
diverse: in technical terminology, they exhibit a high degree
of variance, both genetic and non-genetic.

Unfortunately, ideological warfare between the environmentalists and the hereditarians is still rampant over all the
broad territories of the social sciences, including the domain
of education, with the environmentalists still maintaining a
vigorous offensive campaign. In order to bring the two parties
together, the biologist must point out that their dispute is as
silly and nonsensical as that between the Bigendians and the
Littlendians in *Gulliver's Travels*. The simple and fundamental fact is that neither environment nor heredity is the
more important. Both are necessary: and all characters of all
organisms are the result of their joint effect. When organisms differ, whether children or chrysanthemums, their
differences may be due either to differences in environment
—the conditions in which they have grown and developed—
or to differences in heredity—their genetic outfit of chromosomes and genes. But the share of either party in the combined operation may differ markedly in magnitude. Thus the
share of genetic determination is high in mammalian hair-
and eye-colour, in human colour-blindness and various forms
of mental defect, while that of environmental modification
is high in size in flowering plants, or health and physique in
human beings. One of the most striking effects of the last war
in Britain was the notable improvement in the physique and
mental energy of children, especially in the lower income
brackets, due to the special rations of milk and vitamins that
they were given—and this, of course, has had obvious
implications for their education.

If a diversified population is made genetically uniform,
as when man makes pure strains of dogs or wheat by inbreeding and selection, more and more of its visible variation

will be modificational, due to differences in the environment. Conversely, if its environment is made more uniform, as when an unselected strain of plants is grown in carefully controlled conditions, more and more of any visible variation will be genetic in origin.

All these points are highly relevant to our subject. Thus it is no good training a colour-blind child as a painter or a signalman, or trying to give mental defectives the same educational treatment as normal children.

Perhaps the most surprising recent modificational change in man has been the secular trend towards earlier maturity, summarized by Dr J. M. Tanner in his book, *Education and Physical Growth*, which has been going on in western countries for over a century. The onset of puberty, as measured by the age of girls at menarche, has been advanced by some four months every decade, from over 17 years in 1850 to about 13 years today. This is correlated with a marked acceleration of growth. During every decade of the seventy years since systematic measurement was undertaken, Western adolescents have become bigger and heavier by about one inch in height and four pounds in weight; and maximum stature is now achieved by the age of 16 to 17 in girls and 18 to 19 in boys, as against nearly 25 in 1850. Whatever its cause (which must certainly be sought in environmental factors, such as nutrition and healthy up-bringing), this trend has various interesting consequences for education. Thus boys' schools are faced with a shortage of trebles for their choirs; and an average boy of 15 is likely to be taller and heavier than his woman teacher, and is pretty certain (on the basis of Kinsey's findings) to have stronger sexual urges. And the average girl of 15 is now physiologically and psychologically a woman. The adolescent crisis, with all its problems, has been shifted to an earlier age: and educational systems must adjust themselves to this fact.

Then there are various kinds of genetic difference for the educational system to cope with. There are large quantitative differences in general intelligence, from potential genius to mental defect. There are also large quantitative differences in the genetic rate of mental development, notably between quick learners and rapid developers on the one hand and slow learners and maturers on the other (this, of course, has nothing to do with modificational changes in rate, such as the trend to earlier maturity I have just discussed). More research is needed to determine how far rapid learning is correlated with general genetic intelligence and with high final performance, and whether slow learning may not sometimes give more solid later achievement; but clearly our educational system must take account of these differences.

There are also qualitative differences in the way children's minds handle the raw materials of experience, notably between visualizers, auditory types, concrete manipulators, and abstract verbalizers: the traditional verbal type of education is not very good for visualizers. Then there are differences in creativity and in the capacity for absorbing and regurgitating academic knowledge. The two are not by any means always strongly correlated—indeed, a recent American study has come out with the sweeping assertion that "the class standing of a student has no correlation with his performance in later years", and that the U.S. educational system, based as it is on scholastic achievement, may actually be eliminating some of the best creative talent.

And there are psychosomatic differences, of anatomy and temperament. After Sheldon's work, not even the most aprioristic theorist could expect that a skinny ectomorph would react to the educational process in the same way as a comfortable endomorph with perhaps double the ratio of guts to muscle and a radically different pattern of hormones being squirted into his bloodstream.

The problem is how to be educationally fair to the whole range of these and other variant psychophysical types. Here as in so many fields, the abnormal may shed light on the normal. Let me give two examples of abnormality and how society can deal with it. Myopia—short-sightedness—is mainly genetic in origin: in a form severe enough to be a handicap in ordinary life if uncorrected, it occurs in well over 1 per cent. of human children. It may have been of selective value during early stages of man's history, when fine craftmanship was needed and craftsmen were kept out of the fighting; but in the modern world it is a severe handicap. However, the modern world has found a way to circumvent the genetic handicap, by correcting the defect with spectacles: by this means the abnormal is brought within the range of the normal.

My second example concerns schizophrenia. As mentioned earlier, this grave mental disorder, which all too often leads to certifiable insanity, affects nearly 1 per cent of all human beings, in every country and in every type and level of culture. Recent work is making it highly probable that it has a genetic basis, and is due to a metabolic error which interferes with the working of a very important bit of the brain's machinery for building up the chaos of sensations into an orderly system of perceptions. The confirmed schizophrenic's perceptual world is a disorderly one, not unlike that into which a normal person makes a brief entrance by way of mescalin or lysergic acid.

The educational relevance of this is becoming evident. The genetically schizoprone child begins to manifest overt schizophrenia when the disorderliness of his perceptual world throws him thoroughly out of gear with the ordered "normal" world of his fellow human beings and his society. As one might expect, this usually happens during the adolescent crisis. Psychiatrists are busy devising chemical tests for the schizophrenia-provoking compound in the blood and urine and psychological tests for its effects on the percept-building functions of the brain. These tests could best be carried out in the school. And then the schizoprones could be given special educational treatment designed to correct their defective picture of the world and reconcile it with that of their normal fellows. They need corrective spectacles for their defect of inner vision, as myopic children need them for their defect of external vision.[1] Meanwhile, educationalists have the inspiring possibility of lightening mankind's heavy burden of schizophrenia.

The general challenge of human diversity remains. Man

---

[1] Recent research shows that schizophrenics are many times more prone to suicide than normal people. Most student suicides seem to be correlated with incipient schizophrenia.

is the most diverse and variable of all organisms—variable anatomically and physiologically, intellectually and temperamentally, genetically and environmentally. And a large degree of diversity is a source of strength to human societies, especially to high civilizations. Successful psychosocial evolution demands a variety of gifts, temperaments and talents. A humanist society needs men of action and men of thought; scientists and artists; brain-workers and labourers; saints and policemen; adventurers and stay-at-homes; eccentrics and established civil servants; leaders and led.

It would be a good thing if the numbers of the too abnormal and the too defective could be reduced, those of the more intelligent and more gifted increased; and perhaps one day eugenics will get busy on this. Meanwhile the problem is how to utilize the existing and potential diversity of people to the best advantage of society and of themselves. As Dwight Ingle has written, individuality must be a parameter of the educational process.

The Texan anatomist, Roger Williams, who has done more than anyone else to establish the full range of man's anatomical and physiological diversity, has proposed *Free but Unequal* as the motto of a modern society. Freedom in inequality is a good basis for an educational system to work on: but as a goal for it to work towards, I would suggest *Varied Excellence*.

In any case, educationists must assuredly struggle against conformism and must resist the imposition of all dogmatisms, including their own. They will remember that cultural and individual diversity is precious in itself, and will strive for vivifying variety and against monotonous mediocrity. They will try to ensure that the more gifted children are not bored and frustrated by being kept back to the level of the average, the less gifted not made to suffer by being pushed beyond their capacities. They will try to provide a range of opportunities to meet their pupils' range of aptitudes. But they will hold fast to the humanist vision of variety in unity, and will endeavour to provide a common ground of thought and action, a unitary vision and framework of ideas which all human variants can share.

135

One point deserves special consideration. The increasing complexity of modern societies demands an increasing number of men and women of great ability and high competence to run them. It should be a prime duty of our educationalists to meet this demand. For this, genetics and education must join hands. We need a comprehensive selection system to catch as many potential geniuses and top people as possible, and once caught, we must give them an education designed to help and permit them to realize their capacities to the full. Failure to do this will lead to a running down of national efficiency and national achievement. It will also be a lamentable waste of that most valuable of all human resources, mental and spiritual power, and will prevent many potential geniuses from developing their precious talents.

In doing justice to human variety, educationalists will be accused of encouraging an élite, and of aiding new class differences. That must be faced. Nature is not egalitarian; societies must always be stratified in some way; and, whether you call them an élite or anything else, outstanding people are needed at the top.

Since education sets out to promote the right development of human beings, a proper understanding of the developmental process is important both for educators and for their political masters. Development is a natural process and must be studied as such, not in the abstract or scholastic terms of some presumed intrinsic vital principle or ideal purpose, nor merely in terms of its elements and origins, real or postulated. In regard to human (and animal) behaviour, the reductionists are still very vocal. Orthodox psychoanalysts spend most of their time searching for origins, overzealous behaviourists and neurologists proclaim that mind does not really exist or is merely an epiphenomenal resultant of matter, the Skinnerian school asserts that nothing but learning is of much importance in behaviour.

We must beware of all such "nothing-buttery". Whenever anybody says or implies that something is "nothing but" something else, or is explicable "merely" in terms of its elements or origins, we can be quite sure that he is wrong. In opposition to *reductionism* I suggest *eductionism* as a rallying cry to constructive thought. It means that when we are considering any development process of biological or human development (including of course evolutionary as well as individual development), we must begin with its operative function—its organized end-result and how well or badly it works in the business of life. Then we can analyse it into its elementary components and search for its origins. Finally we must study the operation of the process itself to find out how it *educes* its end-results out of its original components—in more general terms, how it actualizes its potentialities. Only then shall we be in a position to set about improving it.

The first and basic fact about human development is that it is not a mere unfolding of a miniature model; the developing human being passes through a series of radically different stages—infancy, childhood, boy- or girlhood, adolescence, maturity, each of them demanding different educational treatment. Yet many educational systems have insisted on treating the child as a miniature man, and others as so much blank or plastic material, to be written on or moulded at the educator's will.

In reality, the development of man, like that of all other organisms, is what biologists call *epigenetic*. It is a cybernetic process full of feedback mechanisms, and produces both complexity and emergent novelty. The modern science of development has shed the title of Embryology in favour of Epigenetics.

Today we are beginning to explore the mechanisms by which the genetic code inscribed in the chromosomes of the egg is translated, through the co-operative interaction of the genes and their environment, into bodily and mental organization, and are discovering various of the principles at work. One of the most important of these is what Waddington has called the *canalization* of development into a number of channels, each leading in a definite direction towards a specific kind of end-result, while any departure from their "right" course is automatically corrected. They thus have a high capacity for self-regulation—perhaps self-direction is the better term.

Such processes Waddington calls *creodes*, meaning by creode "a pathway of change" which is equilibrated, in the sense that the system tends to return to it after disturbance. Some creodic systems provide alternative but sharply distinct pathways: thus the creode concerned with the development of our reproductive organs can be switched by means of the sex-hormones into one or other of the pathways leading to normal male or to normal female organization. And of course many creodes can be somewhat deviated by the environment without giving an abnormal result: sunlight can deviate our skin-colour from pale to dark, exercise can lead to extra muscular development.

Other creodic canalization systems may be more plastic and permit the organism to wander off in various directions over the developmental landscape. This applies especially to the behaviour-systems of higher vertebrates, whose learning capacities permit them to adjust their activities to their experience; and most notably to that of man. The creodic system concerned with human behaviour is so widely and deeply plastic that it not only provides for an extensive range of wandering and branching in later stages of development, but can be radically influenced at its early roots and in the main trunk of its growth.

It has to integrate competing and even conflicting basic drives and desires into some sort of wholeness. Integration can be prevented by depriving the child of some basic ingredient for his psychosocial development, as John Bowlby has shown for maternal deprivation. It can be traumatically distorted by internal conflict culminating in repression, which then leads to the projection of one's own guilt and repressed aggression outward on to others, or sometimes inward on to oneself. An excessive load of infantile guilt may produce anything from a cruel superego and a harsh and unforgiving morality to an inferiority complex and an over-meticulous conscience. A load of sexual guilt at adolescence coupled with frustrating failure to achieve satisfactory social adjustment may lead to an amoral and asocial bored indifference, to a retreat from the social world into neurotic depression, or at the other extreme to an assertion of the ego's

significance by violent and anti-social actions: meaninglessness is the parent of delinquency. Clearly educational
systems cannot supply psychoanalytic treatment, nor can
they take over from parents the job of looking after the early
development of their babies. But they can do much to correct
infantile distortions, to bypass adolescent frustrations, and to
replace meaninglessness with significance. They can do so by
providing opportunities for fulfilment and satisfaction in all
sorts of ways, and in particular through activities which make
boys and girls feel that they matter, that they mean something to themselves and other people.

In general, educators should try to provide children with
what I may call integrative creodes—developmental processes which will steer themselves towards more effective
integration of knowledge and behaviour. This does not just
mean the "moral education" that conventional educational
systems aspire to give: it means providing children with more
effective systems of canalizing their own moral, intellectual
and spiritual development.

How to achieve this aim properly is a matter for much
research and experiment, though clearly a good deal can be
done to avoid frustration by providing channels of fulfilment
within the educational system. Group projects can obviously
help in giving a feeling of significance. But we want to know
what sort of project and what type and size of group gives
what degree of significance to what kinds of boys and girls.
Already both sociologists and ethologists are busy studying
the general problem of social groups and their working; the
time is ripe for educators to combine with their colleagues
in tackling the special problem of the role of groups in
education.

Ethology is giving a useful lead, by studying the formation
and function of behavioural bonds in animals. In higher
vertebrates, such as geese, jackdaws, monkeys and chimpanzees, such bonds involve emotions: so they do in ourselves,
but in ourselves they can be reinforced by empathy, which
seems to be present only in rudimentary form in animals.
They include the bond between parent and offspring, the
pair-bond tying mated birds together, the gregarious bonds

in bird flocks or caribou herds, the social bonds operating in a wolf pack, rookeries or baboon communities.

Some important points are emerging. First, all behavioural bonds have a genetic built-in basis, but only develop to full effectiveness through some sort of learning under the impact of experience. Secondly, close and intimate relations between individuals usually involve an element of hostility, and therefore of aggression as well as of attraction. Thus in what is often misleadingly called the courtship of birds, the attitudes of the mates as they approach each other are the resultant of a mixture of attraction, attack, and escape. These attitudes have then been (metaphorically) seized upon by natural selection and turned into formal displays serving as moves in the reproductive game. Or, to take an example from a species which (like man) is both social and quarrelsome, fights between jackdaws are prevented from leading to a really serious outcome by means of a special ceremonial attitude. If a younger and weaker jackdaw is in grave danger during a fight, he presents the conspicuous and vulnerable back of his head to his more powerful enemy; and this automatically inhibits the aggressive urges of the attacker.

All such behaviour has been largely ritualized—turned into formal activities of functional value; and my third point is that in this process of ritualization, the element of aggression is made to serve some useful purpose, whether by redirecting it against other objects or by utilizing its energy in the performance of the bonding ceremony. Rituals are thus methods for transforming frustrating conflict situations into biologically significant activities. Clearly, something similar could be achieved in education. Many conflicts could be acted out in ritual form, many activities could be given added significance by putting them in a formal setting.

All primitive societies cope with the adolescent crisis of their young people by elaborate *rites de passage* laden with significance. Can our over-intellectual and over-technological societies not devise something similar? Could not all the adolescents of one age-group be called on to undertake some challenging individual exploit, as with the boys at Gordonstoun? Could not another age-group be called on to embark on group adventures and projects; and yet another to under-take some sort of service to the community? Could not the end of school life be celebrated in a more formal and expres-sive way than now?

The broader question poses itself: cannot the passage to the adult stage be linked with the larger community of the nation and the world, in something which would really deserve the title of a national, or even international, service? The rudiments of such a service are there, in organizations like the Peace Corps, the Organization for Relief and Development, and the Youth Conservation Corps, in exploration societies and travel studies. We should get on with the job of seeing how to link them all up in some flexible and imaginative scheme which would give outlets for many types, would provide some sense of significance and direction to many chaotic lives and would become an import-ant part of the country's educational system.

In this, as in many other ways, our education could be made a more integral process, and the split between its two functions of intellectual instruction and professional training on the one hand and moral education and character-formation on the other, which began over two thousand years ago with the rise of the Sophists in Greece, could be bridged.

*   *   *

Education, in spite of all the hot air expended in lip-service to its importance, still has an unfortunately low status in our competitive technological world. Yet it has a crucial and very special role to play in helping the technological world through its present crisis. If it is to raise its status, the educational profession should devote a great deal of thought and energy to understanding the nature of that crisis and devising methods to meet it. As a result of the knowledge explosion of the last hundred years, the evolutionary process, in the person of post-Darwinian man, is at last becoming conscious of itself; it is time that the educational process, in the person of the educational profession, should become conscious of itself as the essential psychosocial organ for transmitting and transforming human culture.

141

ESSAYS OF A HUMANIST

A first step, it may be suggested, would be the establishment of an Education Council, on a par with the Arts Council, the Medical and Agricultural Research Councils, and the D.S.I.R., though with its own special functions. If the Government were chary of giving it the same financial aid as the other Councils, the educational profession ought to set it up on its own, confident that, once in being, it would speedily prove itself and attract official support.

The overriding question for such a body to consider would be this: how can education help in bringing coherence, significance and direction into our chaotic, fragmented, and bewildered world? It could do so in two distinct but interdependent ways, one primarily concerned with knowledge and the handling of outer experience, the other with personality and the handling of inner experience. The first aims at providing a meaningful picture of the outer world and a coherent pattern of ideas to help in understanding it and our relations with it: The second at promoting the development of a meaningful and coherent inner world, and providing a pattern of activities and opportunities to help in achieving it.

Various projects of this sort are already being initiated. Thus the Gulbenkian Foundation is organizing a series of conferences on education. The *Universities Quarterly* contains a report on proposals made at one of these conferences, for a Social Science Research Council which would tackle the problems of higher education in Britain. The new University of Sussex is hoping to undertake a large-scale research investigation of its own operations. And we have had the provocative Crowther and Newsom Reports, and finally the Robbins Report on Higher Education, whose drastic proposals for extension and reorganization are even more challenging.

An obvious and immediate task is to hammer out the details of a general curriculum which, as I have suggested, would reflect the unitary vision provided by modern science and learning. Sample textbooks should be prepared, and consideration given to teaching aids and to learning through projects. Such a curriculum would bypass the split between Sir Charles Snow's Two Cultures; its very unity would give it a new dynamism, with sciences and humanities reinforcing each other instead of being driven into mutual hostility. Critics like Dr Yudkin claim that this is impossible and would involve "the falsely optimistic idea of an age of Leonardos". On the contrary, I believe that it is perfectly possible to give every normal boy or girl some real understanding of what science and history and literature and industry are about: this could be achieved by the age of 15 or 16, though it would be much easier if secondary education were extended to 17 or 18. And the first year or so of college or university life should supplement necessary specialization with some reasonably advanced study of complementary fields. Throughout, there would be emphasis on the role of concepts and techniques of thought and expression as mental tools and machinery enabling one to comprehend external reality more readily and more satisfactorily; on projects as well as on customary instruction; and on making the process of learning and understanding enjoyable.

In relation with this major theme, one subgroup might study examination systems, with particular reference to devising methods of testing understanding as well as knowledge, and of assessing creativity and general capacity as well as academic performance; another might look into ways of preventing the growth of anti-rational ideas and superstitions; another would deal with the question of teaching aids[1]; and yet another with the problem of ensuring an adequate supply of top-level minds for the nation's business, administration and culture.

The second major study would be concerned with inner development—how the educational system could encourage the growth of integrated personalities, at war neither with

---

[1] Some years ago when I visited the Great Mosque of Kairouan, the oldest university in the Western World, I saw not only the traditional method of instruction, groups of students listening to a learned man by a pillar in the mosque itself, but was also shown with pride the modern improvements, in the shape of little classrooms round the great courtyard, each equipped with a blackboard. I had never before realized the revolutionary impact that this teaching aid must have had on the educational process. I suspect that the impact of our modern teaching aids—films, closed-circuit TV, radio relays, and teaching machines will be even more revolutionary.

themselves nor with society. Such a system would bypass the split between intellectual and moral education. In order to achieve this, the educational process must help in the resolution of inevitable conflicts. It must provide opportunities not only for enjoyment but for fulfilment of many kinds, especially for consolidating the boy's or girl's identity and for achieving a sense of significance. Such a study will be difficult. It will involve calling for help from a whole range of ologies—anthropology, ethology, psychiatry, child psychology, physiology. It will involve facing the problem of evil and what theologians call original sin, and reconciling some violently opposed points of view. But if it is successful it could lead to new conceptions of the functions of education and a revolutionary overhaul of its methods. It could also enable education to affirm its importance as a transformer of society as well as a transmitter of culture; for the products of such a personally fulfilling education would undoubtedly press for the establishment of a real-life Fulfilment Society.

Then there is the exciting territory of non-verbal and non-intellectual education to be explored and exploited.[1] One interesting new project, now being encouraged by Unesco, is the awakening of the critical spirit and the development of standards of taste, by means of what has been called screen education. This aims to introduce children to the audio-visual language of film and television, including its style and its techniques, and to develop their conscious appreciation through discussion and evaluation of what they see.

There is the problem of education's link with the arts. Our educational system is now being called on to supply the nation with large numbers of scientists and technicians; should it not be required to furnish a similar quota of artists, architects, sculptors, decorators and designers? There is the question of giving schoolmasters (and, of course, schoolmistresses) opportunities for research and creative work.

[1] For an imaginative treatment of this subject, see Aldous Huxley's Utopia, *Island*; and for an interesting experiment in art appreciation and art education, in which pupils learn to grasp the "perceptual unity" of a painting or a situation when illuminated for a tenth of a second, see Hoyt Sherman's *Drawing by Seeing*.

There is the tricky problem of tests—intelligence tests, open-ended tests, aptitude tests, personality tests. There is the problem of how far the practice of meditation can replace or supplement formal religious services, prayer periods and scripture lessons.

And there is the world problem. The world has become one *de facto*. It must achieve some unification of thought if it is to avoid disaster (let alone proceed to political unification), and this can only come about with the aid of education. We must remember that two-fifths of the world's adult population—700 million grown men and women—are still illiterate, that the world's provision for education at all levels is lamentably inadequate, and that the underdeveloped countries are all clamorously demanding more and better education.[1] Britain, as an ex-colonialist country and the senior partner in the Commonwealth, has a special responsibility for meeting these demands. Unesco is also doing a great deal in this urgent matter.

Beyond all these and many other particular problems there is the great question-mark of so-called Adult Education. With the combination of more automation, compulsory leisure, and greater spending power, we shall soon be faced with the task of extending our educational system to meet the needs of the entire adult population. Education in the ordinary limited sense seems destined to become only a part of a comprehensive and continuing process. Perhaps we should look forward to the establishment of a National Education Service. In any case, the prospect opens up all manner of exciting possibilities, which it is the duty and the privilege of the educational profession to explore.

Obviously, university Departments of Education will be required to enlarge their functions, and they will become the academic agencies of the country's educational system. By fully establishing the claims of education to equal rank with other subjects in the university curriculum, both as a field

[1] There is some danger of regarding education as a panacea for emergent nations. The cost of the immediate educational facilities demanded is often too high. Attention should also be given to schemes for economic development, which will enable an adequate educational system to be set up eventually.

for research and a course of study as well as a training for a job or a career, they could do much to improve its public image.

But these are matters of detail. For, make no mistake, the basic task before the educational profession today is to study and understand the evolutionary-humanist revolution in all its ramifications, to follow up its educational implications, and to enable as many as possible of the world's growing minds to be illuminated by its new vision of human destiny.

It is a strange and rather disturbing fact that in the one-volume abridgement of Toynbee's monumental *Study of History* there is only one short section on education, dealing merely with the impact of modern democratic theories on the subject. If the educational profession rises boldly and successfully to meet the challenge of the new knowledge and the new vision which it reveals, new histories of mankind will not only devote much more attention to education as a major function of man in society, but will single out our age as the historical moment when education was reorganized as an integral part of the psychosocial process, and became pre-eminent among all the agencies concerned with human destiny.

BIRDS AND SCIENCE[1]

MANY people still seem to consider a pre-occupation with birds as having a sentimental rather than a scientific interest. In point of fact, ornithology is now in the lead in a number of branches of biological research, notably in regard to species-formation and related evolutionary problems, to animal behaviour and psychology, to the study of animal populations, and to the extension or contraction of geographical range.

Much of this scientific lead now held by ornithology has been gained since the end of the war. This is perhaps most striking in the domain of speciation. Birds are the only class of either animal or plant kingdom of which the total number of existing species is known with any approach to accuracy. As Ernst Mayr of Harvard has shown, there are about 8600 species of birds, and it is extremely improbable that more than a dozen or so will ever be added to this number.

Actually, the number of bird species recognized today is nearly 2000 fewer than that of a quarter of a century ago. This apparent paradox is due to the discovery, first fully realized in birds, that most species, at least in all groups of higher animals, are polytypic; that is, they consist of several distinct geographical races or subspecies, which do not overlap but replace each other in different areas of the range of the species. Some are quite distinctive—for instance, the British Pied Wagtail is distinguishable at a glance from its continental relative, the White Wagtail. Many of these were described by the older naturalists as full species, but are now merged in polytypic species. The total number of bird subspecies is about 28,500, an average of 3·3 subspecies per species.

Among birds one finds beautiful demonstrations of species in the making. When a species has extended far round the

---

[1] This chapter is based on two articles on the International Ornithological Congresses held at Uppsala and Basle.

147

world, differentiating into a chain of subspecies in the process, it may happen that the two ends of the chain come to overlap, and then behave as distinct species, being unable to interbreed with each other, although each link in the chain can interbreed with its neighbours. A familiar example is provided by the Lesser Black-back and the Herring Gull, which, where they overlap, as in Western Europe, are to all intents and purposes "good species" (though occasionally isolated hybridization occurs), but are also the two ends of a circumpolar chain of subspecies, extending right round the northern hemisphere.

Then there are the beautifully diagrammatic cases where two parts of an originally single and uniform species have been forced apart by the Ice Age, and later, on the retreat of the ice, have been able to meet again. According to the amount of differentiation that has taken place during their isolation, they may then either behave as two distinct species, like the Marsh and Willow Tits in Britain, or the two Tree-creepers (distinguishable by voice but hardly by eye) in Central Europe; or still as mere sub-species, which then interbreed and produce an intermediate and variable population over a large area; or finally may reveal themselves as on the very brink of full speciation, by hybridizing only along a narrow zone where their ranges meet, the zone being kept narrow by the fact that the hybrid population is genetically unbalanced and therefore not so well adapted to survive as either of the parental forms. This occurs with the Carrion and Hooded Crows in the Old World, with the eastern and western woodpeckers known as Flickers in the U.S.A.

This brings me to my next point—the general realization that almost all the characters that distinguish groups of birds, whether subspecies, species, genera or higher categories, are adaptive. When I was beginning my biological career, the whole notion of adaptation was largely rejected by the avant-garde biologists. Even when the power of natural selection and the consequent general importance of adaptation were admitted, the majority of biologists still clung to the idea that most of the characters that distinguish one species or subspecies from another were neutral or

148

accidental. Today, thanks partly to the development of a mathematical theory of selection, based on modern genetics, but partly to experiment and close observation in the field, largely in birds, the very idea of biological neutrality has been called in question, and all or almost all characters are regarded as either adaptations or else as the correlates or consequences of adaptation.

Some of the best examples of this have emerged from careful studies of the evolutionary effects of competition on closely related species of birds. Thus Dean Amadon of the American Museum of Natural History has studied the differentiation of the birds known as Honey-creepers. These have had a history in the Hawaiian Islands very similar to that shown in the Galapagos archipelago by the famous Ground-finches which finally forced Darwin to abandon the theory of special creation of species in favour of a belief in their gradual evolution. It is clear that, in both cases, the oceanic island group was first colonized accidentally by a small number of individuals of a single species. These, finding little or no competition (for oceanic islands are beyond the range of dispersal of most land birds), not only multiplied but soon began to evolve in different directions when isolated, on different islands, eventually producing a considerable number of distinct types adapted to different ways of life. The particular point brought out by Amadon is this: when two related species, with slight differences in way of life (and, in consequence, in form of bill) are found together on one island, they show a sharper difference than when they are not in competition. Thus, an insect-eating Honeycreeper with a relatively large bill, when found on the same island as one with a small bill, has evolved a more curved and considerably more powerful bill, which enables it to get at insects under bark that are unavailable to its smaller competitor.

The converse of this is seen where the absence of competition has led to the evolution of a less specialized type. Thus, as David Lack of Oxford has shown, in one of the Canary Islands where the Blue Tit is the only member of the genus to have found its way, it has invaded the habitats of other tits, such as the Coal Tit, Great Tit, and Crested Tit, and has taken on various of their characters and habits, so that it is becoming like a common denominator of the whole tit genus.

Next I must mention some of the recent research on bird song. For some time the puzzling fact has been known that some birds do not have to learn their songs, while others will sing something if reared in isolation, but that something is less perfect and less striking than, and often quite different from the normal, which has to be learnt.

It was generally supposed that the nestlings normally learnt their songs from hearing their father singing—and, indeed, that this teaching of the next generation was the, or at least a, biological reason for the continuation of song in many species after the eggs have hatched.

Now, however, Dr Poulsen of Denmark, working with Chaffinches (a species belonging to the second group, in which song is partly innate but partly must be learnt), found that neither in the nestling stage nor even after they were fully grown did hearing the normal song have any effect on the birds. They had to be exposed to it for the comparatively short period next spring when their testes were maturing for the first time and pouring the male sex-hormone into their blood; otherwise they produced only a poorer, more twittering song than normal, and one that always lacked the final flourish. Apparently, this is a sensitive period, like the much shorter one just after hatching in geese, during which alone a certain sign-stimulus can be effective and become imprinted. As with the geese, other comparable sign-stimuli can be substituted for the normal. Thus, if young Chaffinches during this period are kept where they can hear other birds, such as Linnets or Canaries, they will give a passable version of their song, although neither they nor their ancestors have ever heard it. And once the "wrong" song is imprinted, they can never learn the right one, any more than the man-imprinted gosling can ever follow the "right" parent.

Once the sensitive period of a few weeks is over, imprinting is no longer possible, and previously isolated birds are condemned to go on singing the imperfect innate song for the rest of their natural lives.

Poulsen also found that, of two other members of the finch family (Fringillidae), the Reed Bunting's song is wholly innate, but the Linnet's wholly acquired, so that it will not sing at all if kept in auditory isolation. All mere call-notes, by the way, seem always to be fully innate in all species.

Later, with the aid of the sound-spectograph, Thorpe at Cambridge was able to confirm these results, and also to extend them. Whereas Poulsen thought that the learnt part of the song, whether of their own or some other species, could not be imprinted before the young birds themselves were coming into song, in their first year of maturity, Thorpe found that they could learn some time before this. The problem now is to discover the earliest age at which the birds' nervous system has developed this imprinting capacity. Once imprinting has taken place, the form of the song is firmly fixed.

The invention of the sound-spectograph, by the way, is proving of great importance to scientific ornithology. This apparatus can provide a graphic record of a bird's song which is complete as regards both frequency, amplitude, and duration of the notes; further, the record can be played back and reconverted into sound, and can be slowed down to permit the adequate study of rapid passages. The scientific study of sound-patterns by the unaided human ear has always been difficult; but with the help of this apparatus they have become amenable to detailed and quantitative analysis —more so indeed than most visual patterns.

Meanwhile Sauer in Germany had given a rigorous proof of the innate (genetically determined) nature of some very elaborate patterns of behaviour, including song. Sauer's studies well illustrated this fact. He kept Whitethroats isolated from their parents and from all other species in sound-proof rooms throughout life. In spite of this, they produced all their calls and songs, all their food-searching and display actions and sexual behaviour, in completely normal fashion, and in regular sequence. This valuable piece of work constitutes the final defeat of those numerous psychologists and sociologists who persist in contending that there is no such thing as instinct or innate behaviour, and that everything must be in some measure (or even wholly!) learned. Some of them simply shut their eyes to the facts, while others confuse genetically predetermined development with learning by experience.

Sauer also discovered the interesting fact that elderly birds suffering from illness or mere senility will show a regression of behaviour to that of earlier stages. In some cases they will even revert to the food-begging behaviour of fledglings.

A remarkable post-war development has been the enlisting of the interest of the amateur in ornithology. The most notable step in this direction was the founding of the British Trust for Ornithology, which carries out various scientific investigations almost wholly through the efforts of amateurs; there is also the spectacular growth of the Swedish Ornithological Society to nearly 2000 members, mostly amateurs, since its foundation in 1945; and there have been similar developments in the U.S.A., Holland, and elsewhere.

The result is that many men and women who would at best have become isolated amateur bird-watchers or sentimental bird-lovers become interested in the scientific aspects of bird-life, and have the satisfaction of helping in research and the advancement of knowledge.

Many important new facts, for instance on the fluctuation in numbers of particular species (like the Common Heron or the Great Crested Grebe) within a country or region over a period of years, or the relation of clutch-size and of egg and fledgling mortality to geographical position and climatic variation, have been discovered only through such co-operative efforts by amateurs.

I cannot help contrasting the state of ornithology today with that of fifty years ago, when I began to be seriously interested in it. Then, there was a school of old-fashioned ornithologists who maintained that ornithology in a country like Britain was played out and that there was nothing more of importance to discover; and in opposition to them, a group of younger-minded men who realized that in every field, except mere descriptions and records of dates and localities, scientific ornithology had barely been born. The questions of subspecies and species-formation, of sexual selection, of behaviour in general, of accurate studies of migration by means of ringing, of the size and fluctuations of bird populations, of homing—these and many others were just beginning to appear as exciting and profitable fields for research.

Today in all these fields ornithology is reaping a valuable crop of results, and has the satisfaction of feeling that it has shouldered its way up from a neglected and relatively un-important subject to being an essential and often a leading branch of biological science.

Elsewhere in this volume I have given some account of the remarkable progress made since the war in ethology, as the scientific study of animal and human behaviour is now called. Here I will only make the general point that the comparative study of behaviour is showing that birds and other higher vertebrates have in the course of evolution been equipped with a definite set of behaviour organs as well as a definite set of physiological organs. Indeed, we can now in principle analyse the entire complex of a bird's behaviour into the workings and interactions of a quite small number of behaviour organs, just as was done in the classical system of comparative anatomy based on visible gross structures such as the skeleton or the blood-vessels, which laid the first foundation for the later idea of evolutionary relationship.

Of late years, Poulsen of Denmark has turned his attention from song to anting—that curious action seen in so many passerines (but in no other group), when a bird picks up ants and then proceeds to rub its head and beak against the under-side of its wing and sometimes against its tail feathers. Some observers have thought that this was a substitute for a bath, others that the formic acid from the ants served to kill or discourage bird-lice and other parasites. However, to the sorrow of the parasitologists, Poulsen seems to have estab-lished that the action is a purely instinctive one, released by any irritating substance. In normal life it is carried out by the numerous passerine species which eat ants. It has been suggested that this is to get rid of the irritation caused to the sensitive areas of the bird's face by the formic acid squirted out by the ants as a defence. Later studies have shown that the birds may be agreeably stimulated by the ants and other irritants. Maurice Burton's tame Rook even spread its wings eagerly over burning matches or flaming paper.[1]

Now let me pass to another subject—the present change in world climate. It may seem curious that ornithology should be playing a leading role in this problem, but in fact birds are not only extremely sensitive indicators of climatic change, but also readily visible and intensively studied. Thus, in Iceland during the past fifty years, as Finnur Gudmunsson reported, while the average temperature has been rising and the glaciers receding, the number of regularly breeding bird species has increased by almost 10 per cent. Many birds have extended their range from the south to the north, and the island is on the verge of losing its only truly Arctic breeding species, the Little Auk, because the pack-ice round which it finds food for its young is now too far from the coast.

Sometimes, indeed, the birds may give pointers to the meteorologists. Dr Peitzmeier, for instance, reported that in north-west Germany during the last eighty years a slow decrease in bird numbers for most of the period was followed by a rapid increase later, especially in the last decade. Some birds, like the Hoopoe, actually disappeared entirely from the area, to reappear later. This applied almost entirely to migrants arriving late in the summer; other species showed a steady increase and expansion of range (at the rate of about three miles per annum in the Stock-dove) during the whole period. A painstaking analysis of the weather records showed that these facts were apparently to be correlated with June mean temperature, which showed a fall followed by a sharp rise. This would affect the late-arriving migrants. Earlier spring months did not show the fall, or to a negligible extent.

[1] Mammals may show similar behaviour. Thus one species of Tarsier (a primitive primate) is named *Carbonarius* because it picks up embers from camp-fires. Dr Oakley (*Man*, 1961) suggests that this primate propensity may eventually have led on to man's deliberate use of fire.

Dr Merikallio spends most of his time walking the length and breadth of Finland taking bird censuses, with the result that Finland is the only country for which a first approximation has been made of the numbers of all species of breeding birds. He revealed an even subtler indicator of climatic change. Two closely-related species of the finch genus *Fringilla* breed in Finland, the Brambling, which occurs as a winter migrant in Britain, and the common Chaffinch. In the south of Finland only Chaffinches are found: in the north, only Bramblings. In the intermediate areas, as demonstrated by elaborate sample counts carried on over many years, the ratio of breeding Chaffinches to breeding Bramblings steadily falls from south to north. There is of course a zone where the ratio is unity and the two birds are present in equal numbers: this zone has moved some 200 kilometres northward in the last thirty years, and the whole gradation has shifted in favour of the Chaffinch, the more southerly species.

Just the same has happened with the two species of Titmouse with overlapping ranges in Finland, the more southern Crested Tit and the more northern Siberian Tit; and also, as Dr Salomonsen has shown, with the boundary between two non-overlapping subspecies of finch in Greenland, the Mealy Redpoll and Hornemann's Redpoll. And all these large shifts of an objectively observable zone of demarcation can be connected with climatic change.

However, the most spectacular extension of range known in biology, that of the Fulmar Petrel, can have no connection with climatic change, since the extension has been southward instead of northward. This species, which originally did not breed outside the Arctic and a few outlying islands like Iceland and St Kilda, has in less than a century colonized Faeroe, Shetland and other Scottish islands, and almost the whole of the coastline of Britain, with parts of Norway, too. James Fisher's careful analysis has shown that in all probability the spread has been due to man providing this almost omnivorous species, with its very plastic food habits, with new sources of food, first by whaling, and later by trawling where the fish are gutted at sea. This new food is provided in lower latitudes than the Fulmar's natural food, and the

155

species has colonized a vast new area of sea and coastline as a parasite of man.

Another method of extension of range is shown by the Fieldfare. This Scandinavian thrush never bred in any part of America until about twelve years ago, since when it has become quite common in one area of south-west Greenland. Dr Salomonsen's careful analysis showed that this was due in the first instance to abnormal weather conditions in January 1937, when a large flock of the birds endeavouring to move south-west from southern Norway to Britain was caught in a violent south-east gale and whisked away to Greenland. Here they found a birch-scrub area where, thanks to the general improvement of climate, they were able to breed. A curious point is that, while in Scandinavia they migrate in winter, in their new home they have ceased to be migratory and stay in the same place all the year round.

Biologists have always assumed that accidental dispersal must have played its part in the geographical distribution of animals, but here is an example of the process actually happening, and in this case extending the range of a species to a new continent.

Some of the most original and exciting post-war work is that of Dr Kramer, of Wilhelmshafen, who studied the direction of flight during migration—on captive birds! He discovered that during the migration period Starlings kept in a suitable cage would perch or flutter towards one particular direction—that which they would have taken if they had been free. But, more than that, by an ingenious arrangement of mirrors he proved that the direction of the sun's light determines the direction chosen. By manipulating mirrors at a number of openings round the cage, he could make birds which normally gathered towards the SW., gather at the opposite or NE. side, or half-way round, towards the NW. Here at last is a core of hard fact in the welter of speculations as to how birds find their way, a known natural phenomenon instead of a mysterious "sense of direction" or a hypothetical reliance on magnetism.

This work has since been extended and deepened in many ways. All day-migrants appear to navigate by the sun; for

156

this, they are equipped by heredity with a wonderful built-in mechanism to enable them to evaluate both solar direction and azimuth and also an internal clock to take account of time.

Even more extraordinary, night-migrants navigate by the stars. This was proved by placing birds in a planetarium during the season of migration and noting in which direction they gathered. When the pattern of the night sky was rotated by 90°, the birds too shifted their direction. This, by the way, appears to be the first time that a planetarium has been employed for purposes of scientific research.

Meanwhile, Kenneth Williamson has demonstrated the importance of wind-drift as a factor in migration. When migrating birds (particularly night-migrants) meet with strong winds, they may be drifted hundreds of miles out of their course, especially as they tend to turn down-wind and ride the drift. It is chiefly owing to this "drift-riding" that we have the huge autumnal passage migration down the east coast of Britain. The birds' "true" migration route is down the west coast of Continental Europe; but large numbers of birds are drifted across the North Sea and utilize its eastern coast as guiding-line for their further southward journey. The method by which the famous table-top mountains of Venezuela (which were taken by Conan Doyle as the site of his Lost World) received their bird population has long been a puzzle. Mayr and Phelps favour the idea that colonization by wind-drift has played a major part.

Among many interesting studies on particular species I may mention Sladen's work on Adelie Penguins. In this species, as in many other penguins, the members of a pair indulge in a joint ceremony of bowing and billing at the nest; and it was assumed that this was essentially a mutual display, serving as a psychological stimulus and an emotional bond between the members of a mated pair, such as occurs in grebes, herons and many other forms in which both sexes share in incubation and looking after the young. In the Adelie Penguin, however, Sladen found that the ceremony might take place between a half-grown chick and its parent, or between two chicks. Its function, he thinks, is largely one of recognition, or rather the visible and emotional confirma-

157

tion of recognition. Thus it is given when one of a pair
returns from several days at sea to relieve its mate on the nest,
and equally when a chick recognizes its parent returning with
food. It is clear that we shall have to take this idea, of the
recognitional value of display, into serious consideration.

Sladen also shed new light on the gatherings of half-
grown young that are a feature of their species. It used to be
supposed that these were veritable crèches, in which the
young were looked after by a certain number of adult guards.
It now turns out that these "guards" are not guards at all,
but merely non-breeders or birds which have lost their eggs
or young. They are impelled to stake out territories and build
large stone nests for themselves—all, of course, quite use-
lessly. The chicks gather round them, but they do nothing
to minimize their reproductive failure—they neither protect
nor feed the young birds.

Guy Mountfort's intensive study of that curious (and
curiously little-known) bird the Hawfinch includes many
new facts, notably concerning its courtship, which is exceedingly protracted. This is due to the aggressive character of
the unmated female, who drives away the courting male
until—sometimes only after two months—his advances and
displays succeed in breaking down her "dominant" attitude.
During this long refractory period, the cock's behaviour
shows an extraordinary mixture of fear and desire, sometimes rising to a veritable paroxysm of ambivalence. Part of
the cock's display is an elaboration of the submissive attitude
in which the grey patch on the vulnerable nape is prominently exhibited, just as Lorenz found in Jackdaws. This may
alternate, or be combined, with the innate aggressive attitude.

I must also mention the Brown Owl, blind from hatching
with cataract, which learnt to find its way about a room with
other birds in it by reacting to the sounds and air-currents
which they made; and the brilliant comparative studies of
Professor Portmann of Basle, on the proportionate size of
the cerebral hemispheres in relation to the rest of the avian
brain, which enables us to place different groups of birds
in an evolutionary hierarchy with reference to the development of intelligence.

In a very different field we have Dr Vogt's remarkable investigations of the famous guano-producing cormorants of Peru (about eleven millions of them!). I have space to mention only two of his points. One is that each bird produces annually several times its own weight of dry guano. The other is Vogt's finding that the occasional "bad years", when millions of the birds starve to death, are due to abnormal weather conditions further south, off Chile, which change the normal direction of the winds and force the fertile food-bearing cold waters of the Humboldt current out westwards beyond the powers of the birds to reach. Here is an excellent example of the precariousness of the balance with the environment in which so many species live.

I myself had made an extensive study of morphism (or genetic polymorphism) in birds, meaning by that the coexistence, within a single interbreeding population, of two or more sharply distinct forms or phases, the least abundant of which is present in numbers too great to be due merely to recurrent mutation (as with albinism or haemophilia in man). R. A. Fisher, our great English geneticist, pointed out over thirty years ago that such a state of affairs could not exist except as the result of a selective balance. If, for simplicity's sake, we take an example where there are only two distinct forms or morphs—as, for instance, the presence of a considerable percentage of so-called bridled or spectacled individuals in the Guillemot population of Scotland and Iceland—each must have both some selective advantage and also some selective disadvantage vis-à-vis the other, though in this and many other cases we do not know in what the advantage or disadvantage may consist.

I was concerned to demonstrate how widespread morphism is among birds, and how it constitutes a valuable but hitherto little-studied method of intra-specific differentiation, the different distinct morphs adapting the species to a corresponding set of distinct environmental conditions. A classical case is the polymorphic mimicry of Cuckoos' eggs, adapting the species to a wide range of different hosts; but there is abundance of other examples, such as the multiplicity of forms with different display-characters in the male Ruff, the partial migration of many passerines, or the variations in clutch-size in many species whose nestlings depend on insect food. In this last case the existence of a strain laying fewer eggs is an insurance against bad seasons, when there is not enough food to go round, and large broods therefore suffer more than small ones. There are species which make more than one distinct kind of nest, like the Brush Turkeys studied by H. J. Frith in the Australian region; and others, like some of Darwin's Finches (Geospizids) from the Galapagos, in which a certain number of males breed in immature plumage, the percentage varying markedly from island to island. Often the relative numbers of the different morphs vary geographically. Thus the percentage of bridled Guillemots increases with a combination of increasing north latitude and humidity; the proportion of dark to light Arctic Skuas grows progressively greater to the northward.

Another interesting but puzzling fact is that only a minority of bird orders or families include many examples of visible morphism, for instance, herons, skuas, petrels and owls, while ducks, grebes, divers and gulls do not. I concluded my report with a plea for more intensive study of the problem, especially the possible existence of "invisible" morphisms in birds, such as the blood-group and sensory-threshold morphisms of man and the chromosomal (inversion) morphism so widespread in fruit-flies (Drosophila); and a reminder that every case of morphism is a challenge to the biologist and naturalist.

I must also mention Tratz's film of vultures (*Gyps fulvus*) in Austria. The film is not particularly good, but the fact that every summer two hundred or so vultures come to a restricted area in the mountains near Salzburg, some 350 miles from their nearest regular haunts, is of great interest.

Finally, the International Ornithological Congress at Basle in 1954 was notable for the reappearance of delegates from Russia, after sixteen years of absenteeism. Four Russian ornithologists came and read papers, including Dementiev, whom I last saw at the celebrations of the Academy of Sciences in Moscow in 1945. We then discussed the possibility of his undertaking a *Handbook of the Birds of the*

U.S.S.R., a project to which he had been stimulated by seeing the great *Handbook of British Birds*, published during the war. This has now been published, and the work constitutes a very valuable addition to our ornithological knowledge.

THE COTO DOÑANA

"SQUACCO?" asks James Ferguson-Lees, the tall assistant editor of *British Birds*. "Two on the Sopiton lagoon," replies Max Nicholson, editor-in-chief of that same periodical. James enters the particulars under Squacco Heron and proceeds through the entire local bird-list, ending with House Sparrow and Tree Sparrow at the tail end of the Passerines.

This intensive scientific ritual takes almost an hour and a half. It is gone through every evening by the group of ornithologists assembled in southern Spain on the Coto Doñana, at the edge of the Marismas west of the Guadalquivir estuary.

The Coto Doñana, together with the neighbouring Coto to the westward, extending to near Huelva (where Christopher Columbus visited the friendly Friar who successfully interested Madrid on his behalf), and with the swamps or marismas to the north, is the largest roadless area in Western Europe today. A Coto, I should explain, originally meant an area where hunting and shooting rights were in private hands. These two Cotos, comprising over 125 square miles, make a great nature reserve. There are a few villages on the edge of the marismas, but in the Cotos themselves the only habitations are scattered cottages and huts and the Palacio de Doñana, where our party lives. (A palacio in Spain, by the way, is no more a palace than a château in France is a castle: it means a large country-house.)

The area has been a place of pilgrimage for ornithologists ever since Abel Chapman rented the Palacio for several years running at the turn of the century. His various books are a sad tale of bird-slaughter, but also a mine of information. Here Witherby, the king-pin of twentieth-century British ornithology, as a young man collected a wide range of species and laid the foundation for his knowledge of European birds. And in 1952, Guy Mountfort came here with Roger Peterson of American Bird Guide fame to glean information for the Field Guide to the Birds of Europe, in which the two of them were co-operating.

Out of this visit grew the idea of three expeditions to make an intensive study of the wild life, notably the birds, and the general ecology of the region. The present party consists of twelve members. Besides Guy Mountfort, James Ferguson-Lees and Max Nicholson, it includes Eric Hosking, prince of bird photographers, George Shannon, dentist and nature cinematographer, Phil Hollom, export-import agent and co-author of the *Birds of Britain and Europe*, Valverde, an excellent young Spanish naturalist, Lord Alanbrooke, and some others.

The wild life of the Coto is extraordinary. Out of the windows of the Palacio you can always see Godwits and Stilts on the marismas, and often a herd of twenty or thirty Fallow Deer, tawny and spotted, making their way across the edge of the marshes, with a few Red Deer in the background. The great eucalyptus trees round the Palacio are tenanted in the mornings by groups of brilliant Bee-eaters, those most beautifully tailored of birds, sailing round now and again before re-alighting. Great Spotted Cuckoos perch and call there, and in one of them a Barn Owl has its nest. The Palacio itself is tenanted by a number of Starlings—not the vulgar Common Starling but the Spotless Starling, of an immaculate black with blue sheen: it replaces the Common Starling in parts of the Mediterranean. Swallows of course are abundant, and so are Goldfinches. At night Stone Curlews give their haunting call and you can hear the Common Nightjars churring and the Red-necked Nightjars making a queer wooden clappering. (The difference between the two "songs" is doubtless due to the need for distinctiveness when two closely allied species inhabit the same area.)

Wild Boars abound. One rarely sees them by day, but the evidences of their activities are everywhere—large areas of soil rooted up in search of roots and tubers: Yellow Iris corms are specially sought after.

There are Foxes, Hares, and Rabbits, and to my surprise Lynxes. Two other even more surprising animals are the

163

Mongoose and the Genet, both of them African species which in the prehistoric past have somehow managed to cross the strait of Gibraltar into Europe. The Genet, a large spotted civet-like creature as big as a big cat, is even found in southern France.

Among other mammals are Hedgehogs, Edible Dormice, and Badgers. There are a fair number of snakes, a terrapin, a water-tortoise, and a good many nice land-tortoises (it gives a north European quite a shock to come on one of these armour-plated creatures plodding through the cistus); geckos in the house, a chameleon, several ordinary kinds of lizard, one of them a foot long and green, and also a queer short-tailed type with much reduced legs, and a legless and eyeless subterranean species; frogs and toads, and a beautiful little green tree-frog.

The birds demand a section to themselves: our expedition has identified nearly two hundred species. Here I will only record the wonder and pleasure, for an Englishman, of seeing so many large and conspicuous birds which have elsewhere in Europe been killed off or crowded out by man. Grey Herons, Purple Herons, Squacco Herons (an inconspicuous buff when at rest, but flashing white wings when they rise), Night Herons, Cattle Egrets, their white tinged with pale buff, and the fine snow-white Little Egrets, whose evening flocks winging their way to their roosts are a miracle of beauty. We have also seen the biggest or at least the heaviest land bird in Europe, the Great Bustard—not on the Coto, but in the rolling wheat-lands north of Jerez—magnificent creatures, alertly scanning the horizon for danger.

Flocks of Godwits are everywhere in flight, looking like great white-patterned moths with outstretched proboscis; Curlews, their bills even more proboscis-like against the sky; Avocets in the lagoons, pure white and black, sweeping their upcurved beaks from side to side over the muddy bottom in search of food. There are Sandgrouse in the dry places, Ducks and Grebes of many kinds on the lagoons, with Coots and Moorhens, and in the tall reed-beds that splendid creature the Purple Gallinule, as heavy as a big fowl, blue-purple with turquoise breast and bright red bill and legs.

As a climax there are the birds of prey—three kinds of Vulture, four Eagles, two Kites, two Harriers, two Buzzards, the grand Hobby Falcon, Kestrels, and others. Hosking wanted to get pictures of Vultures feeding, and a cow which had died was hauled out into the scrub in the early afternoon. But when we went that same evening to put up a hide, the entire carcass had been devoured. A later hide was more successful, and some fine photographs were secured of these great birds, so magnificent when soaring aloft, so revolting close at hand, with mean head and naked, snaky neck—snaky to reach out into a corpse, naked to avoid contamination of feathers by putrescent carrion. Like all soaring birds, they depend on air-current rising from the earth as the sun warms it up, so they are not visible in the sky until the morning is well advanced, and on cold, dull days we looked for them in vain.

We saw their full magnificence of flight at Arcos, twenty miles east of Jerez, on the river Guadalete. The river has carved its way into the hill on which the town is built, leaving a great cliff dropping vertically down from the town's edge. In a recess on its face, the Vultures come every afternoon to roost. From the plaza we watched them planing down one after another on their nine-foot spread of wing, sometimes banking and turning for another approach, sometimes dropping their under-carriage and disappearing to roost under their chosen overhang.

Magnificent though the Vultures were, for me the greatest attraction of Arcos was the swarm of Lesser Kestrels, over two hundred of them, which haunted its cliffs. These beautiful birds are smaller than our Common Kestrel, but much more striking to the vivid russet and blue-grey of their colouration, and much more exciting in flight. The Common Kestrel is capable of hovering with its head fixed in one point of air, adjusting itself even to strong and gusty winds by movements of its tail and bastard wings; but this miracle of control is outdone by the Lesser Kestrel's aerobatics. They swoop in towards the cliff face like projectiles, are lifted as

ESSAYS OF A HUMANIST

in an invisible elevator by the updraught, pursue each other round the buttresses in chattering arrowy flight, circle up and out, interweave and swoop in again.

The abundance of wild life on the Coto recalled earth's virgin days. For me it brought back memories of the game plains of East Africa, though there the birds of prey are of different species, and the deer are replaced by antelopes.

The human life of the place, too, was a survival. Most of the families scattered over the Coto Doñana lived much as their ancestors would have done in medieval times, except that they have a tractor available for certain operations. The Palacio imposes its feudal hand on the whole. It exudes an atmosphere of princely hunts; indeed every Spanish king since the early seventeenth century has come there to shoot and hunt and stick pigs. Its apogee must have been reached in the shooting parties of the Edwardian epoch, commemorated in an array of faded photographs filling the living-room and spilling out along the passages. These show aristocratic shooting picnics with the guests seated on folding chairs; beach picnics with a carriage for the ladies; noblemen, guns on shoulder, wearing striped shaps as protection against the prickly scrub; stags being gralloched; game-keepers leading horses laden with deer carcasses à la Millais; proud sportsmen with their bags of ducks, their lynxes, their heads and horns, their wild boars; culminating in a large signed photograph of Alfonso XIII standing by the head of a fine royal stag and dated 20 January 1924.

During this period it had been equipped with *le confort moderne*, in the shape of two bathrooms (one enormous one in the royal suite) and W.C.s. But even so, as we discovered after enquiring the reason for the excessive irregularity of meal-times, there was not a single clock or watch among all its inhabitants and dependents!

Chief among these are the three brothers Chico. Antonio is the head keeper or *guarda*—a fine figure of a man with strong features and an intimate knowledge of the wild life. Rafael is our majordomo: he insists on doing all the serving at table, always with a toothpick in his mouth, and not allowing any of the maids to help, though this delays matters

166

considerably. He finds our habit of eating eggs for breakfast very un-Spanish. The third and rather uninteresting brother, Curro, comes along on all our expeditions, carrying the food and heavy apparatus, content to wait in somnolent vacancy while we do our explorations.

Then there is Pepe, nicknamed *El Mono*, the monkey, an extraordinary character: smallish, but powerful and very agile, moving like a feline, and like a wild animal too in hating to sleep in a house. Whenever an eagle's or a kite's nest is found, it is Pepe who climbs the tree to report on its contents. He is up in a flash, embracing the trunk with his arms and pressing with his bare feet, whose skin is like ancient linoleum. Whenever a rabbit is sighted, he sets his dogs after it with passionate cries and whistles, and often joins in the chase himself. His small mobile face is framed in black side-burns, and his deep-set eyes gleam like burning carbuncles. When we were leaving, there was a stretch of mud between us and the launch, and Pepe had to carry us over it. I shall always remember his delight when he was told to hoist my wife on his back, and his amazing feat of carrying twelve-stone George Shannon across the mud and depositing him safely in the boat.

The Palacio Doñana now belongs to the Gonzales family, main proprietors of the famous Gonzalez-Byass sherry and brandy firm. In their great bodegas, we saw the unique Gonzalez-Byass mice. The old night watchman (a man of cask-like build, recently deceased) beguiled his watches by putting out scraps for the mice in the bodega. Then he had the idea of offering them sherry. Each night he put out a glassful with a little mouse-size ladder leading up to its rim. They soon learnt to use the ladder and developed (as who would not?) a marked taste for the tipple. The old ones, it seems, have pretty strong heads, but sometimes the youngsters will finish up reeling drunk. The attendant put out a glass full of sherry and set up the little ladder, and placed a chair for me about four feet away. Within two minutes a couple of mice were licking the base of the glass, and a little later one climbed the ladder and started imbibing. The second mouse soon started up the ladder and eventually managed to elbow (if that is the proper word for murines) the first-comer to one side so that he, too, could get at the drink. Never shall I forget the sight of two little mice indulging this unnatural but delicious appetite.

*   *   *

Near the Palacio, a pinewood was occupied by carboneros —charcoal-burners. Their ancient trade was universal in Europe throughout the Middle Ages, and only died out in more northerly countries with the advent of cheap coal for coking. Like much else from the European past, it has survived in Spain, and it was fascinating to see the work going forward as it would have done in the Weald of Kent and Surrey until less than two hundred years ago.

The men, after first piling up a circular dome of sand, stack several layers of wood, in the shape of trimmed pine-branches, against its sides and cover the whole (except for a strip at the bottom to permit draught) with a thatch of pine-needles. They build a little stairway of sticks up the dome, and make a vent halfway up, then fire the wood, and you see blue smoke curling out of the vent.

When the burning is over, the thatch and the central dome are removed, and nothing is left but a ring of black charcoal sticks on the ground, later to be packed in gunny sacks and carted off to market.

There are also gauchos on the Coto. Like all cowboys, they are proud of their own way of life and of their get-up with stiff-brimmed hats, wide leather shaps, and primitive lances. A few days ago there was a round-up of calves. In this culturally shrinking world, it was surprising to find that the customary method of roping animals had not reached the Coto. Instead of a lasso, the gauchos used a long pole with a running noose at its end to slip over the heads of the calves. The process took two or three times as long as the roping of a steer by an American cowboy and was much less entertaining and skilful.

*   *   *

Our primary photographic target was the Imperial Eagle, of which less than 100 pairs still survive in the world. On the afternoon we arrived, we prolonged our four hours' ride by another hour, in spite of increasing stiffness, to inspect a nest that the guarda had discovered. We could hardly believe our good fortune, for it was in quite a small, low pine, and contained a fine eaglet about two weeks old, already with vicious-looking beak and talons contrasting with its innocent white down. The next three mornings were devoted to the erection of a hide on a pylon of tubular steel scaffolding. The four men working on the precarious structure looked like figures from one of Fernand Leger's paintings of acrobats. Finally the photographers were installed, and were rewarded for hours of heat and cramped waiting by a close view of the imposing parents, which they were able to record on an excellent film—the first to be made of this rare and glorious creature.

Great Spotted Cuckoos were everywhere on the Coto, largish, with a crest on their heads, and a rather harsh rattling note instead of the eponymous cuck-oo which has given our common Cuckoo its name.

Cuckoos with their parasitic habits pose some of the most interesting problems in ornithology, and the Great Spotted Cuckoo is no exception. The common Cuckoo, though as big as a Sparrowhawk, is adapted to parasitize small song-birds. Its egg is only about half the size to be expected from the dimensions of the bird that lays it, scarcely larger than the eggs of its hosts; it also mimics their colour and pattern, as a further safeguard against ejection. Indeed, the common Cuckoo is broken up into a series of races, or *gentes* as the specialists call them, each adapted to one main foster-parent and laying eggs of the appropriate colour: blue for Redstarts, speckled brown for Meadow Pipits, white with black scribble markings for Buntings, and so on. As the parasite is so much bigger than its hosts, the young Cuckoo must be assured of the maximum amount of food. Accordingly it is furnished with an instinct to eject any other object from the nest, such objects being normally the eggs or young of the host.

But the Greater Spotted Cuckoo parasitizes Magpies— birds of its own size. Accordingly its eggs are no smaller

169

than they should be, and up to five or six may be found in one host nest. But in spite of all our efforts, we were unable to discover how the young magpies are got rid of.

The interesting fact has recently been established that many of our small birds will attack stuffed dummy Cuckoos, not because cuckoos look like hawks as has sometimes been suggested, but because the song-bird hosts are equipped by heredity with the instinct to attack this particular enemy of their species. (In America, where parasitic cuckoos are absent, the small song-birds lack this instinct.) This was confirmed by our expedition, which established the fact that Great Spotted Cuckoo dummies are attacked by their particular host species, the Magpie.

* * *

The wind changed to the east on May 11th, and on the 12th the Coto was full of migrants—Woodchat Shrikes perched prominently every few score yards, flocks of Turtle Doves; and seven species of Warbler were seen in a short walk. So on the 13th we set off early for the seashore. The lagoons of Santa Olalla were crowded with Black Terns in gentle moth-like flight. Questing with them we saw, to our great excitement, two White-winged Black Terns. These are the most beautiful creatures imaginable, with a strange and lovely pattern of black, white and dark smoke-grey on their graceful bodies. Normally inhabitants of the great Dobrudja swamps at the mouth of the Danube and the regions to the north and east, they must have been blown right out of their proper range, until they saw and recognized kindred creatures, with whom they then foregathered.

The shores of the lagoon were swarming with newly arrived shorebirds—Common Sandpipers, teetering and calling just as on a Highland Loch; Grey Plover on their way to their sub-arctic tundras, beautiful with their silver-grey upper parts and black breast and belly; Redshank in musically calling flocks.

Then an hour's hot walk to the beach, across two ranges of sand-dunes over which no vehicle can pass. The sand-dunes, by the way, are slowly moving inland. On one trip, I was able to emulate Baron Munchausen in half of one of his exploits, by hitching my horse to the topmost branches of a stone-pine, the rest of which had been buried by the dunes' advance.

On the shore stood groups of strange fish-traps, looking like transmogrified dressmakers' dummies. The beach itself was bordered with a row of what we first took to be fishermen's huts, but turned out to be conical stacks of marram grass, to be used for fodder and bedding for animals. Since marram is the only agency that can hold bare sand, uprooting it obviously prevents the dunes from becoming stabilized, and Max Nicholson told us that Queen Elizabeth issued edicts forbidding this anti-conservationist practice in England. Either her writ did not run in Ireland, or else she did not bother about that difficult and rebellious country, for the practice still continues there, as it does here in Spain.

Now the experts really got busy; batteries of field-glasses were focused on every bush, and deep arguments took place, for the identification of many of the warblers is extremely difficult when they are not singing. Here were Whitethroats, Chiffchaffs and Garden Warblers, perhaps on their way to England; Bonelli's Warblers, Marsh Warblers, Rufous Warblers (handsome and agreeably easy to identify), Cetti's Warblers, Olivaceous Warblers, Melodious Warblers in quantity, Redstarts, Tawny Pipits, Spotted Flycatchers, Pied Flycatchers and various unidentifiables.

One other new arrival demands special mention—a Roller in magnificent plumage of intense deep blue, revealing turquoise on the wings and warm buff on the back when in flight. In Europe, Rollers have a curious distribution, extending poleward to south-east Sweden, but avoiding the moist Atlantic climate of similar (and lower) latitude to the west.

*   *   *

Meanwhile my wife had been acting as botanical collector, and had got together some 250 species, duly identified later by Kew. Her find of the Marsh St John's Wort, familiar in marshy spots in England, represented a new locality for the species. It was lovely to see the wild Lupins, the cerise Gladiolus and the blue Spanish Iris rising out of the heathland, the masses of papery white and yellow Cistuses, and the local Snowflake (Leucojum) growing in sandy dry soil. Then there were strange plants like Urginea, whose huge bulbs are used to make rat-bane (not penicillin, as one of the keepers gravely informed us!) and yet are eaten by wild boars; giant Thrifts, deep-rooted and growing in bare sand; blue Jacob's Ladder in the open tracks; an occasional Arbutus; and a single Wild Olive. The Cork Oak is one of the main trees: the local people make their beehives out of slabs of its corky bark.

*   *   *

When I was a boy, bird-watching was a pleasant pastime: today it has become a science, even though mainly carried on by amateurs. My most enduring recollection of the Coto is of the zealous young men extracting the last ounce of ornithological information from the assembled party after supper in the Palacio, a speck of scientific activity in the primitive expanse of the marismas.

RICHES OF WILD AFRICA

THE political issues of Africa are now under the limelight in the centre of the world's stage. But while Africa resounds with the speeches of politicians, white and black, professional and amateur, and our newspapers are full of their printed echoes, another African issue is approaching its crisis—that of wild nature.

The issue is this—whether Africa's wild life and wild nature can survive, or whether they will be destroyed or whittled down to a poor remnant by the rising tides of over-population and industrial and other "development".

In the long run, this is more important than the political issue. Politics we shall always have with us: but if wild life is destroyed, it is gone for ever, and if it is seriously reduced, its restoration will be a lengthy and expensive business. Furthermore, wild life and wild nature constitute one of the major resources on which a politically evolving Africa can rely. In this chapter, I shall try to justify these statements and elucidate the issue.

In 1961, my wife and I came back from the most interesting assignment I have ever had—to report to Unesco on the Conservation of Wild Life and Natural Habitats in Central and East Africa. This meant a three-month journey involving ten countries, twenty-five National Parks or comparable areas, and discussions with well over a hundred administrators, experts, scientists and politicians, both white and black.

Our first impression was of the wonderful riches provided by nature in the region; our second of the many and multiform dangers threatening these riches; and our third of the new hope of averting these dangers.

There are many kinds of riches in the world. A country can enjoy mineral or commercial or architectural riches; it can be rich in its history or in its modern achievements. The unique riches of Africa are its wild life and wild lands—

173

tropical rain-forest and tawny thorn-bush, great lakes and rivers and waterfalls, rift valleys, volcanoes, strange plants and trees, and the fantastic variety of animal life, from butter-flies and fish to great reptiles, birds, and above all mammals —for all the world to wonder at and enjoy.

Strictly speaking I should have said Eastern Inter-tropical Africa. Although South Africa has killed off almost all its large mammals, it may be included as an annexe because of its National Parks and its unique and marvellous flora. But the vast Congo is essentially a forest region, whose wild life is less outstanding and much less readily seen or studied; North Africa belongs to the Mediterranean region, not to the real Africa; and the riches of West Africa consist largely in its unique art and in its success in retaining the flamboyant survivals of barbarian pagan kingdoms and black Islamic feudalism within the framework of modern in-dependent nationhood

The region I am dealing with includes East Africa as its kernel, with Somalia and the southern part of the Sudan and Ethopia to the north, a narrow strip of the Congo to the west, Mozambique, Angola, and the Rhodesias and Nyasa-land to the south, and a southern fringe extending across from the north of South Africa through Bechuanaland and the Kalahari to the coast of South-West Africa.

Let me enumerate its main natural riches. First, geo-graphical and geological. Two great rivers, Zambesi and upper Nile, one with the most stupendous, the other with the most extraordinary waterfall in the world; the world's second largest lake; the staggering Rift Valley system, with its unique rift lakes, its volcanic phenomena, and its great fault-block, upthrust to form Ruwenzori, the Mountains of the Moon: hundreds of volcanoes, in all stages of activ-ity and grand decay, including the most majestic of all, Kili-manjaro, and the best African mountaineering peak, Mt. Kenya.

In East Africa at Olduvai, Dr Leakey has discovered the site of the earliest known ancestral ape-man, Zinjanthropus, together with the primitive tools he made, and has exposed a rock section displaying the entire million-year history of

the Pleistocene, with its changing climates, its evolving
fauna, and its developing human cultures.

Wild life includes plants, for they are alive as well as
animals. Botanically, South Africa is uniquely rich with its
array of proteas, heathers, aloes, everlastings, and cycads,
some of which extend far northwards into eastern Africa and
even Ethiopia.

Our region also has the strelitzias, strange relatives of the
bananas, the only flowers to be pollinated by birds' feet, its
glorious gloriosas, its splendid scarlet erythrinas, its orchids,
its capers and impala lilies, its gladioli and moreas, its extraordinary baobabs, its varied array of acacias, not to mention
the only African member of the cactus family. On its mountain slopes there are bush begonias, giant nettles and dense
stands of bamboo, wild olive and pencil cedar or giant
juniper; and on its mountain uplands, moors of alchemilla,
tree heathers and tree hypericums, and the fantastic giant
lobelias and senecios.

But the vertebrate animals, with the mammals as their
central core, constitute Africa's main riches. The variety of
eastern African mammals is astonishing, and so are often
their numbers. There is still an abundance of relatively easily
visible creatures—elephants, hippos, warthogs, rhinos,
giraffes, lions, leopards, servals, cheetahs, hyenas, zebras,
buffaloes and baboons, monkeys and mongooses, hyraxes
and hares, and a unique array of antelopes large and small—
eland, hartebeest, topi, oryx, sable, roan, gnu (wildebcest),
kudu, waterbuck and gerenuk, through lechwe, gazelles,
bushbuck, reedbuck, impala, steinbuck and klipspringer to
the little duikers and tiny dikdik.

And the sight of great herds of topi or gnu or zebra
galloping across the open plains, of a troop of elephants
coming down to drink and play, of a pride of lions on a kill,
of sausage-like hippos in and out of the water, of a herd of
impala leaping in all directions, of prehistory incarnate in a
rhinoceros, of a family of giraffes cantering along like
elongated rocking horses—any of these is unforgettable, a
unique contribution to the riches of our experience.

Besides these, there are many less frequently seen but

175

wonderfully interesting mammals—chimpanzees and gorillas, bongos and situtungas, bushpigs and giant forest hogs,
wild dogs and bat-eared foxes, otters and wildcats, civets and
genets, polecats and honey-badgers, furred mole-rats and
naked sand-rats, elephant-shrews and bush-babies, porcu-
pines and pangolins, springhares and squirrels, and the
strange nocturnal aardvarks and aardwolves.

African bird life, if not quite so unique, is certainly
glorious. Nowhere else can one see such an array of large and
conspicuous birds—ostrich, the largest of them all, secretary
birds and shoebills, bustards and guinea-fowl, francolins and
quail, cranes, herons and egrets, ibises and storks, spoonbills
and openbills, sand-grouse and painted snipe, coursers and
pratincoles, darters, lilytrotters, skimmers, finfoots, tree
ducks and comb ducks, pigmy geese and plovers, avocets and
stilts, gallinules and crakes, glossy and superb starlings,
hornbills and kingfishers, sunbirds, rollers, bee-eaters,
parrots, coucals and touracous, pigeons and doves, vultures,
eagles and many other birds of prey, and, most amazing
single spectacle, half a million or more flamingoes all together.

Then for the more ornithological birdwatcher, there are
the proliferating families of weavers, oxpeckers and whydah
birds, shrikes and mousebirds, flycatchers and barbets, wax-
bills and white-eyes, and finally the surprising sight of
familiar British or European species, some resident in Africa
like stonechat and dabchick, but mostly on migration like
willow warbler and yellow wagtail, common and wood sandpipers, stork and swallow.

Among reptiles we have a widely diversified array of
chameleons, an abundance of large and easily visible crocodiles, the splendid monitors, and a great variety of other
lizards and of snakes. African amphibia, on the other hand,
are chiefly noteworthy for a negative reason—the absence of
all newts and salamanders.

Among fishes there is one of the three surviving lungfish,
Protopterus—a far more interesting "living fossil" than the
much-publicized coelacanth. There are also the cichlids (including the edible and exportable Tilapias) which in the

great African lakes show a unique evolutionary proliferation into dozens of endemic species and genera: and on the East African coast—a fact which is often forgotten—there are fine coral reefs with splendid coral-fishes, as well as mangroves and mudflats with masses of the strange amphibious fish *Periophthalmus*, the mudskipper.

Among land invertebrates we have the termites, whose existence has promoted the evolution of such strange creatures as aardvark and pangolin; whose cellulose-digesting capacities play a vital role in the region's metabolic turnover; and whose nests have a decisive effect on its landscape and ecology: the many ants, from the blind militaristic columns of driver ants to the symbiotic species living in the galls provided by the gall-acacias: the terrifying wild bees, which provide the African with his main source of sugar and the honey-guide bird with the wax on which he uniquely feeds: the lake-fly whose asemblies of many million individuals form huge clouds drifting across the great lakes: the super-polymorphic swallowtail, *Papilio dardanus*, whose females exist in many wholly different forms and patterns, several of them close mimics of unrelated butterfly species: the stick-insects and praying mantises.

And I must mention the tsetse fly, for though by virtue of its capacity to transmit trypanosomes it is a scourge of man and his domestic animals, by the same token it has saved many areas for wild animals.

This rich variety of wild life and nature constitutes a major resource only in what I may call Africa's Wild Life Area— the drier eastern and southern parts of sub-Saharan Africa. Their total extent may be estimated at about 3,250,000 square miles out of the continent's total of 11,250,000: this is a little larger than the United States before Alaska's statehood.

At least a third of the region consists of poor-quality marginal lands. Most of these are ecologically very brittle habitats, liable to deteriorate and lose productivity if cultivated or in the least degree overgrazed. When overgrazing is severe, as in much of the Masai and Samburu areas, the country rapidly degenerates towards desert.

The area on which I was able to report—the three territories of the Central African Federation, the four East African Territories, Portuguese East Africa, a couple of National Parks in the Republic of South Africa, plus Ruanda-Urundi and a narrow eastern strip of the Belgian Congo on which I was able to get adequate information, amounts to just over 1,500,000 square miles—a little more than the size of Europe without the U.S.S.R.

There are still a great many wild animals left in this Wild Life region of Africa. But they are spread over a very large area; their distribution has become increasingly patchy, and their overall numbers have been grievously reduced. A century ago South Africa harboured tens of millions of large mammals: today they survive in any density only in a few National Parks and Reserves.

Many parts of Kenya and Tanganyika and the Rhodesias which fifty years ago were swarming with game are now bare of all large wild life. Throughout the area, cultivation is extending, native cattle are multiplying at the expense of wild animals, poaching is becoming heavier and more organized, forests are being cut down or destroyed, means are being found to prevent cattle suffering from tsetse-borne diseases, large areas are being over-grazed and degenerating into semi-desert, and above and behind all this, the human population is inexorably mounting, to press ever harder on the limited land space.

In addition, there are many who believe that with the inevitable advent of African governments in most of the territories, game will be regarded as so much meat conveniently provided on the hoof (it is worth recalling that in Swahili and various Bantu languages there is only one word for "wild animal" and "meat"), that National Parks will be looked on as unwanted relics of "colonialism" or as a silly European invention, of no value to the up-and-coming African States, and that no large wild animals will be allowed to survive in African nature.

Thus at first sight the prospect for Africa's wild life looks pretty bleak, and there are plenty of Jeremiahs who are content to prophesy its inevitable extinction, or at best its

178

restriction to a few small National Parks or other super-zoos.

In reality, matters are not by any means as bad as this. For one thing, the old essentially negative notion of preservation has been largely replaced by the positive concept of conservation. Then there has been a recognition of the importance of scientific study of Africa's wild lands and their ecology. As a result of this there has been quite a new assessment of the value of wild life, both to separate African nations and to the world at large. It has been demonstrated that large areas of eastern Africa will yield more meat (and can yield more profit) via game than via cattle or cultivation; while the wild life conserved in National Parks is proving to be a major source of revenue from the tourist trade.

The Africans (or some of them) are beginning to realize the value of wild life for themselves, as a source of much-needed meat, of revenue from visitors, game licences, or ivory, or finally of national pride and international prestige. And finally, international agencies like F.A.O., Unesco and I.U.C.N. (the International Union for the Conservation of Nature) and various Foundations are interesting themselves in the question, so that Africa's wild life is starting to play a role on the international stage.

Perhaps the revaluation of African wild life could be summed up (*sloganized*, I suppose Americans would say) in the phrase "Profit, Protein, Pride and Prestige", with "Interest" and "Enjoyment" thrown in for good measure.

WILD PROTEIN

The wild lands of Africa and the wild life that they carry are a major asset, a natural resource waiting for proper utilization. This resource and the possibility of exploiting it have often been discussed during the last decade, but its full value has only recently been established by scientific study.

This new work has been done mainly by American Fulbright Scholars in Southern Rhodesia and East Africa, but British ecologists like Fraser Darling in Northern Rhodesia and Kenya have also made considerable contributions.

They have conclusively demonstrated the following

179

important points. First, that over much of Africa's wild land a given area will produce a larger weight of animal protein—meat—and might be made to yield a larger financial profit, if managed for game-cropping—that is to say by killing surplus wild animals for meat or hides—than through the medium of cattle or any domestic stock.

Secondly, they have shown that over a considerable further area of land, the presence of a certain number of wild animals may well improve its stock-carrying capacity instead of impairing it. And thirdly and conversely, that much of this large area of wild land is liable to run down-hill and deteriorate if used as grazing for African domestic stock, especially if used without the strictest precautions against overstocking. This last conclusion will not be disputed by anyone who has seen the effect of native stock—mainly cattle, but with some sheep and goats—on the Masai country on either side of the Kenya-Tanganyika border, or the Samburu country in the Northern Frontier District of Kenya. Here large stretches of land are being rapidly reduced to dusty semi-deserts, and if something is not done, will soon become full deserts. Quite simply, the cattle are destroying the habitat.

Thane Riney, the distinguished American Fulbright Scholar, now in Southern Rhodesia but with experience in the United States and New Zealand, probably knows more about this problem than anyone else. I asked him what proportion of Southern Rhodesian land he thought could be more profitably managed to yield game meat than for domestic stock, and his cautious estimate was one-third.

I should imagine that a similar figure would hold for Northern Rhodesia, Tanganyika and Kenya. One-third of these territories would amount to some third of a million square miles—a tidy bit of country! And certainly considerable parts of Bechuanaland, South-West Africa, Mozambique, Angola and (to a lesser extent) Uganda, would fall into the same category.

Dassman and Mossman, two other Fulbright Scholars, have shown that on certain Southern Rhodesian ranches the average cash value of each zebra or wildebeest in terms of biltong and hide is about £8, and of each impala over £4;

kudu, waterbuck, steenbuck and other antelopes could often
be profitably cropped. In one large ranch they estimated the
extra annual income to be derived from game-cropping at
£12,000; furthermore, here and in many other cases, good
management would much increase the game carried by the
land without detriment to the cattle. On another ranch of
135,000 acres they concluded that a game-cropping scheme,
after deducting all costs, including the salary of a European
game manager, could yield a net profit of £8400. If the hides
and other by-products could be marketed, the profit would
be correspondingly higher. On lands with a carrying capacity
of one cow per thirty acres, game can yield more meat than
cattle—about 4 lb. per acre as against 3 lb. per acre. This
yield could be considerably increased if some money were
spent on development to improve the game habitat.

What is basic is the biomass of a habitat, the weight of
mammalian life, per unit area. Here are some figures, in
lbs. per square mile, for different habitats and different
types of animal. Wild ungulates on East African game-plains
without domestic stock: 70,000-100,000 lbs. Wild ungulates
and domestic stock together on East African tribal grazing
lands: 30,000 lbs. plus. Domestic stock alone on tribal
grazing lands: 11,200 to 16,000 lbs. Domestic stock on
good grazing ranges in Western U.S.A.: 26,700 lbs. All
mammals in a West African forest: well under 1000 lbs.
Lee Talbot found in addition that wild ungulate populations
in eastern Africa have a higher rate of annual increase than
domestic stock.

These findings are already having some practical effects.
For instance in 1961 it was announced that Southern
Rhodesia was maintaining about two million acres as wild
habitat, with the hope of eventually cropping the game on it
for meat.

How, you may ask, is this better performance of wild life
possible? A specific reason is that much of this huge area is
infested by tsetse fly: to make it possible to run cattle on it
demands very large expenditure, either for getting rid of the
tsetse by bush clearance or spraying, or for regular inoculation with chemical preventatives of tsetse-borne disease.

But there is also a general ecological explanation, which Fraser Darling has clearly set forth in his book *Wild Life in an African Territory*. The primary reason here is the ecological brittleness of most African wild lands; only a few of them (like the White Highlands and the Kikuyu Reserve in Kenya) are capable of being intensively cultivated or grazed, as are the rich soil areas so largely available in the north temperate zone. When this fact is disregarded (as in the notorious ground-nuts project in Tanganyika), disaster follows. In most areas, cultivation or pastoralism leads to progressive down-grading, with reduction of production or carrying capacity.

The further reason is twofold. It lies in the advantage of the wild life and in the disadvantage of domestic stock, in this type of environment. The wild life is part of a tightly knit natural community which has achieved an optimum balance as a result of millions of years of adaptive evolution. In modern ecological terms, the community has achieved a maximum energy-flow, converting the basic resources of sunlight, soil and water into protoplasm in the most efficient way. It exhibits an ecological division of labour. Its animal component may contain as many as twenty species of herbivore in one area, ranging from huge elephants to tiny dikdik, and including up to fifteen species of antelope, each occupying a slightly different ecological niche. Its herbivorous members utilize the vegetational resources in the most economical fashion—there are grazers and browsers and grazer-browsers and browser-grazers; each species prefers a particular set of foodplants; different browsers browse at different heights, from giraffes reaching to the acacia tops to duikers on the low bushes and herbs; different grazers eat grasses of different length and coarseness; some species also utilize bark or fruits.

Hippo and lechwe promote the conversion cycle in swamps and lake margins by fertilizing the water with their excretions, and so act as intermediaries in converting the vegetational resources of the habitat into fish protein. Elephants are unique, in utilizing an amazingly wide range of habitat, and by combining the functions of aerating the soil, opening up trackways through forest, providing water to the entire animal community by digging, and making fruits available to other species such as impala and baboons by knocking them down off the trees.

Meanwhile the carnivores see to it that the herbivores do not over-multiply and ruin the habitat (as happened in America with the Kaibab deer north of the Grand Canyon when all the mountain lions were killed off). Carnivores are often stigmatised as "natural enemies" of the herbivores. This is wholly false: they are the natural regulating mechanism of the community, and keep the herbivores from getting out of balance with the habitat and destroying their own means of support.

Natural selection has also achieved a balance between wild animals and their parasites, which at worst become tolerated nuisances instead of causing disablement or death. The most striking example of this mutual adaptation is the fact that infection with tsetse-borne trypanosomes produces no ill effects in wild African mammals, whereas it causes the fatal nagana disease in domestic cattle and horses.

The superiority of wild game over domestic stock as a method of utilizing the resources of the habitat is very pronounced in the vast areas, estimated by Barton Worthington to comprise at least a quarter of the entire African continent, which are infested by tsetse, and where cattle therefore cannot be kept. It is possible to render such areas tsetse-free, and various methods have been and are being employed for this purpose—selective bush clearance, spraying, and even game-slaughter aimed at eliminating the vectors of the trypanosome parasite. This last method was tried on a large scale in Southern Rhodesia before the last war, but there is now general agreement that it is not merely destructive of a valuable resource but often inefficient, since various small species may survive and continue to act as vectors. Furthermore, even if successful in the short term, it is of no permanent value unless there is planned settlement and the cleared land is used properly in such a way as to keep the habitat unfit for recolonization by tsetse. Spraying and bush clearance may be successful in enabling African stock to be grazed in the area, but they are expensive, and the area continues to need attention. In any case their use may lead to the introduction of African cattle into areas which could more profitably be reserved for game-cropping. The same applies to the treatment of cattle with trypanosomicidal injections; this will prevent them from developing nagana disease, but may lead to their occupying habitats to which they are ill-adapted.

Cattle, on the other hand, are only grazers, and only one type of grazer at that, and they are subject to all sorts of diseases and disabilities from which the wild animals are free. Their human masters see to it that their natural checks, the carnivores, are not allowed to operate at full strength, and African pastoralists prize numbers of beasts above quality or profit. As a result, native domestic stock is over-increasing, and in many areas is over-grazing its habitat and pushing it downhill towards desert. In the northern half of the Mara River area in Kenya, I have seen with my own eyes a stretch of country which has been thus converted from a splendid wild life area to barren semi-desert in the space of eleven years.

In other places, like the Virunga mountains in the Congo and the slopes of the famous Ngorongoro crater in Tanganyika, the pastoralists are pushing up into a forest habitat and destroying it—and of course also its animal inhabitants, like the mountain gorilla in the Congo.

The African pastoralist tends to look on wild game as competing with his stock for the available grazing, and as the numbers of his cattle increase, he begins to demand the reduction or exclusion of the game. This has already had serious effects in Masailand and the Northern Frontier District of Kenya.

Luckily, ecological study has now demonstrated that in many cases a reasonable number of wild grazers may in the long run improve the grazing for cattle. This is another example of ecological balance: the wild animals keep down certain types of vegetation which the cattle will not touch, and so prevent it from spreading at the expense of the species needed by the cattle. One of the hopeful features in the present situation is that some of the Masai are beginning to understand this.

Furthermore, by utilizing nature's principle of the ecological division of labour, on some types of land man may be
able to obtain a higher yield from a combination of stock and
game than from either alone. In Southern Rhodesia, a combination of eland and cattle is being successfully used, and
Dassman and Mossman have shown that other species like
impala and zebra could be equally useful.

Meanwhile in South Africa, as I discovered to my surprise, there is a rising demand among farmers for certain
types of game on their farms or ranches. Accordingly various
national park authorities, and even municipalities, maintain
considerable areas as wild life farms, and dispatch consign-
ments of blesbok and springbok and other wild game to the
farmers, who make a nice profit out of the meat, usually as
biltong, and the hides. In the Transvaal alone more than
2000 farms or ranches have been stocked in this way. Zebra,
eland, impala, and other antelopes have proved similarly
profitable on Rhodesian ranches.

Game-cropping is already being officially practised in
East Africa. It has usually been started owing to the unfore-
seen over-multiplication of some wild species. Thus the
hippos in and adjacent to the Queen Elizabeth National Park
in Uganda, apparently thanks to the protection afforded
them, recently multiplied to the point where they were
destroying their own habitat and food supply, and the
authorities were forced to take action. Now the hippos outside the park are being cropped to the tune of 3000 tons of
meat a year, and the habitat is beginning to revive.

A similar situation has developed in south-eastern Kenya
with elephants. The African elephant is a highly successful
and resilient species. The elephant population in the Tsavo
National Park has over-increased, probably because they
were debarred from their traditional migrations, and its spill-
over is now being cropped in the Waliangulu region to the
south. The Waliangulu were originally a hunting tribe, and
the cropping is being done by selected hunters from their
ranks, thus providing an outlet for their traditional pro-
pensities. Meanwhile arrangements have been made to
preserve and sell the surplus meat, and it is hoped that the

Government will return to the tribal council most of the proceeds from the sale of the ivory.

The pattern is being repeated in the Bunyoro region of north-western Uganda: here a careful survey has shown that progressive over-population of elephants can be avoided only by killing something over 1000 head annually, and a cropping scheme will, I understand, shortly be put into operation.

In all such cases, the meat is used primarily to feed the local villagers, then distributed to meet the needs of the rest of the tribe, and any surplus preserved or processed and sold to other Africans, as far as possible for the benefit of the local African District Council.

This is a matter of great importance, both physiologically, socially and politically. The bulk of inter-tropical Africa, including most of what I have called its Wild Life Area, is meat-hungry, badly short of animal protein, and therefore below par in mental and physical energy. (In Ghana and other parts of West Africa giant land-snails are a staple article of diet, and snail-farming schemes are being projected.) Game-cropping, it seems, could become a major means for overcoming this dietary deficiency and putting the Africans on a proper plane of nutritional health.

Meanwhile the protein-shortage and consequent meat-hunger is a potent encouragement to poaching. Poaching in eastern Africa is not a matter of individual Africans setting out to kill a buck or two for their own use. It is a highly profitable, and often a highly organized criminal activity. The meat-hungry African is willing to pay high prices for meat, from whatever type of animal and however obtained. I was told that in the Copper Belt game-meat (all poached by gangs using bicycles and lorries) would fetch nearly as much as prime steak in London. As a result, poaching has now in many places become a well-organized racket: perhaps we should say that non-legalized hunting has become a large unregulated industry.

Poaching does provide Africans with some badly needed protein. But because it is illegal and unregulated, it does a great deal of harm, and is in many ways evil. The Reserves and National Parks provide a ready source of game, and the poachers are only too ready to take advantage of it. One of the means they use is to start bush-fires to drive the game out of cover, in the process doing grave damage to the habitat as well as needlessly destroying numbers of small mammals, birds, reptiles and insects. In addition, as a result of what Fraser Darling calls the African's excessive pyromania, many permissible fires that should have been extinguished are allowed or even encouraged to spread.

Then, the poachers employ illegal, cruel and wasteful methods of catching animals. Among these, the wire noose attached to a heavy log is perhaps the most horrible. For big animals up to elephants wire cable is used. Often the snares are not adequately inspected. The wretched creatures drag helplessly at the log, their wounds fester, and they die in agony with flies swarming on their putrefying flesh.

The age-old and almost equally cruel method of pit-traps, usually with stakes at the bottom, is also used, in spite of the hard labour it involves (poaching must be very profitable to induce Africans to undertake the arduous labour of digging a series of deep pits). The use of Acocanthera poison for arrows is also widespread (except, for some unexplained reason, in Uganda). The poison is so deadly that stringent precautions have to be taken in its preparation. This method is especially productive in the dry season and during droughts. If a band of poachers armed with quiverfuls of poisoned arrows hides near a water-hole it can take heavy toll of the game as it comes down to drink.

In Kenya, Africans are not permitted guns, but elsewhere much use is made of muzzle-loaders These are supposed to serve for the defence of crops against wild marauders like bushpigs and certain kinds of antelope, but in fact are to a large extent used for poaching pure and simple In Tanganyika last year there were 70,000 registered muzzle-loaders (registration costs 5s.—a negligible sum in relation to what a buck's carcass will bring in); and perhaps an equal number of unregistered ones.

All these methods, and especially snares and pits, are non-selective and therefore horribly wasteful. They kill young animals and breeding females as well as adult males.

187

Meanwhile Game Departments and National Parks authorities are notoriously under-staffed, both absolutely and also
relatively to other comparable government departments: they
simply have not enough men to cope with poaching. Little
wonder that wild life is diminishing except in a few well-
guarded areas.

It is not only African meat-hunger that makes poaching
profitable. It is also Asian superstition and European taste
for "curios". Indians and Chinese believe (on the basis of
purely magical reasoning akin to that which led medieval
herbalists to the doctrine of signatures) that rhinoceros horn
is a potent aphrodisiac: they believe it so firmly that it now
fetches an extremely high price per pound—much more than
the best ivory; in consequence rhinos are being poached out
of existence, except where well protected.

It is estimated that the total of rhino killed annually in
Kenya alone is 700 to 900, and of elephants 4000 to 5000.
Though East African elephants seem able to hold their own
in a remarkable way, the much smaller and less biologically
resourceful rhino population cannot possibly support such
a drain for more than a few decades at most.

Many giraffes are slaughtered by poachers merely to sell
their tails for fly-whisks; many Colobus monkeys to make
rugs out of their lovely black-and-white fur; many elephants
to satisfy the demands of white men for ivory ornaments,
usually of low aesthetic value, or even for umbrella stands.

Not only mammals are being illegally killed. Crocodiles
are being mercilessly shot for their skins. In some regions
they have been almost or quite exterminated, and in many
others so much reduced in numbers and size that they are
hardly worth hunting: the hunters and poachers are killing
the reptile that grows the golden skin. Furthermore, Cott
suggests that the killing-off of crocodiles may in some cases
actually damage the fishing industry. When very young they
live largely on giant water-bugs and other enemies of fish-
fry, and when they turn to fish-eating in their middle age
the fish they eat are mainly predators of edible species.

Luckily, crocodile poachers can often be spotted, as they
hunt by night with the aid of powerful torches: as a result,

crocodile poaching has been pretty well stamped out in places like the Murchison Falls National Park, though in wilder and more extensive crocodile habitats it still continues almost unchecked.

Poaching is thus a serious threat to wild life in Eastern Africa: some gloomy prophets believe that it will mean the end of big game in the region. However, I do not believe we need be so pessimistic. If extensive game-cropping schemes can be set up, whether the cropping is done by licensed local tribesmen, or Game Department staff, or selected white hunters or their clients on safari, this will go far to satisfy the Africans' legitimate meat-hunger, and to remove the financial and psychological incentives to poaching, both organized and individual. It will also help them to realize that African wild life is a major resource. In the interim, money will have to be found for a really adequate anti-poaching staff in National Parks and Game Departments.

Eastern Africa still has the chance of planning its future scientifically: that is, ecologically. If it is not to lose this chance, it must speedily make a general land-use survey, as a basis for land-use policy and planning. Presumably this will have to be undertaken by the different territories and governments separately, but it will be a great advantage if the various surveys can be co-ordinated on the basis of some common plans and principles. A prerequisite for such plans is to get rid of the assumption that agricultural or other artificial development of natural habitats is necessarily always good and inherently desirable. It is natural for Europeans and other inhabitants of temperate lands with good soils to make this assumption, and it certainly applies to favoured areas of the region, such as the Kenya Highlands. But it is not true for the marginal habitats of eastern Africa, or for most of the equatorial rain-forest area.

The primary purpose of such land-use surveys would be to define three main categories of land. Firstly, land to be allocated to development, including agriculture, industrial uses, and urban development. Secondly, land to be conserved and managed as wild habitat, either for game-cropping schemes; for game-viewing in parks and reserves; for soil conservation and watershed protection, as in certain types of forest; for timber, as in other forests; or simply to be kept undeveloped in reserve. Thirdly, land in which the aim is a profitable symbiosis of wild and domestic stock, as in areas of African pastoralism like Masailand.

In the first category of lands, wild animals are in general destructive, and should be either kept down or—in the case of certain species, like lions, rhinos, and elephants—exterminated, or alternatively given a shot of tranquillizer and removed to a protected area. In the second category, the wild community should be conserved, and the wild habitat scientifically managed. In the third category, a balance between wild and domestic species should be established: the primary aim here will be to educate the African pastoralists and to devise schemes for which they will have responsibility and from which they will derive benefit.

Thirty years ago, on my first visit, there were no National Parks or comparable reserves concerned with the conservation of wild nature, in Central or East Africa, and only two in what I have called the African Wild Life Region as a whole—the Kruger Park in the Republic and the Parc National Albert in what was then the Belgian Congo.

Today in the region as a whole there are nineteen, including the territories of Central and East Africa no fewer than fourteen, together with eight or ten other areas which will eventually deserve national park status.

These twenty-five or thirty reserves are the central redoubt of African wild life and its living show-window to the world. If they were to disappear, the territories in which they lie would lose a large amount of revenue and an immense amount of prestige; and the world be the poorer for the disappearance of a unique source of interest, wonder and enjoyment.

To see large animals going about their natural business in their own natural way, assured and unafraid, is one of the most exciting and moving experiences in the world, comparable with the sight of a noble building or the hearing of a great symphony or mass. A processional frieze of antelopes moving across the African horizon rivals any theatrical spectacle. And I can testify that my first sight of a troop of elephants, old and young, making their way down to water to drink and bathe and play, and then composedly and majestically disappearing into the bush, was profoundly satisfying. I may add that, however many times we were privileged to see wild elephants again, we continued to be enthralled by them.

Nowhere today except in African national parks can human beings enjoy such spectacles: nowhere else can you see grotesque warthogs and bounding gazelles at close range, herds of buffalo and crowds of basking crocodiles, statuesque sable antelopes and unafraid giraffes; certainly nowhere else could you even hope to find tranquil and inoffensive rhinos, witness vast herds of wildebeest and zebra streaming across the plains or watch a pride of lions on a kill from ten yards away. And quite soon, if the pressure of population keeps up, nowhere else in Africa will men be able to walk through un-ravaged montane forest, see the fantastic flora of the open slopes above it, enjoy the solitude and coolness of the high plateaus or the adventures of equatorial mountaineering.

But there is more to African wild life than these imme-diate wonders. A distinguished American vertebrate palaeon-tologist recently told me that he badly needed to visit Africa in order to find out what the animals whose bony remains he had to study really looked like and how they lived. It is even more necessary for the ecologist not only to visit Africa but to work there. Today only in Africa (and to an increasing extent only in the larger African national parks) is the climax of vertebrate terrestrial evolution easily seen and accessible to study.

Theodore Roosevelt on his famous safari in East Africa felt that he was in a pliocene landscape. Though he exagger-ated the time-scale he was right in principle. On the African game-plains we were able to see and study a slice of the pleistocene, before man had discovered how to modify and even control his natural environment, before he had destroyed most of the large animal inhabitants of the habitats he was so ruthlessly and short-sightedly exploiting.

Here we are privileged to be looking at the last remaining

191

portions of the final climax and finest manifestation of pre-human evolution which (unlike the equally rich rain-forest community) are readily accessible for observation and study.

The tropical rain-forest is another original climax community which badly needs scientific study before it is further invaded and destroyed by the assaults of technological man. Though it contains a much smaller array of large vertebrates, it is the most complex ecological system in the world's history, with an overwhelming variety of plants and small animals, notably insects but also vertebrates.

The Canadian north can still show immense herds of caribou, but that is only one species. The North American prairies recently supported stupendous herds of large mammals: but these comprised few species—mostly bison and pronghorn—and in any case they now survive only as protected relics. The Pampas and other plains of South America once carried a fauna as amazing and as varied as that of eastern Africa. But for some unknown reason the great majority of its members became extinct before man got to work on them.

In Africa's wild lands we have a surviving sector of the rich natural world as it was before the rise of modern man, the most conspicuous and easily accessible climax of the prehuman phase of evolution. But the ecological system which it embodies is not merely to be studied as an historical survival, however interesting: it needs investigation in practical detail, for it provides the foundation on which any enduring human ecosystem in the region must be built.

Thus the national parks of Africa are a valuable and irreplaceable world asset, as a source of enjoyment, of interest and of knowledge for the continuing human species as a whole. But they are also an important economic asset of the territories in which they lie, for they attract tourists (our tiresome word for modern pilgrims and travellers) and tourists bring in a substantial revenue, not merely through disbursements in the national parks themselves, but chiefly through what they spend in the country at large, on transport, accommodation, equipment, photography and souvenirs.

The national parks and reserves are now the main reason for tourists coming to East Africa, and one of the main reasons for them coming to South Africa and Mozambique: and the same could soon hold for Central Africa. Tourism is increasing in volume throughout Africa's wild life area, and in Kenya, for instance, has already become the second largest source of annual revenue, to the tune of over £8 million. What is more, it is capable of a very large further increase in the near future (of course, always provided that there is no World War, and no major political trouble in eastern Africa).

So long as Western prosperity continues, with populations increasing and industrialization being intensified, more and more people will want to escape farther and farther from its results, in the shape of over-large and overcrowded cities, smog, noise, boring routine, and over-mechanization of life. Air travel will certainly become cheaper and more popular, and will take more people farther afield.

I would prophesy that the revenue to be derived from tourism in East Africa (which already runs to around £12 million) could certainly be increased fivefold, and probably tenfold, in the next ten years, provided that the business is properly organized. It will be necessary to improve access to the national parks and accommodation within them, catering both for those who like comfort and those who prefer a more do-it-yourself holiday; to provide museums and guides; to maintain a really adequate staff of wardens, scientists and game-scouts, good fencing and anti-poaching measures; and to produce the right sort of publicity.

It will be necessary to work in with travel and tourist agencies in arranging good holidays, efficient itineraries and so-called package tours, to set up new national parks, and to open up new areas, such as parts of the coast and the uplands, for the enjoyment of visitors. And all this without overcrowding the national parks, which would spell their ruination (already at some seasons some of the roads in the Kruger Park are overcrowded with cars full of tourists).

This will cost money—quite a lot of money. So far, game departments and national parks have been the Cinderellas among government departments in Africa: how can they be

193

provided with this extra finance? I would hazard the guess that if the three African territories would co-operate in the matter, and make reasoned application to some United Nations agency for a loan to develop their tourist trade, they would have a good chance of getting it.

Meanwhile the whole question of Africa's wild life and wild lands has been aired on the international stage. I.U.C.N. (the International Union for the Conservation of Nature) in co-operation with C.T.C.A. (the Commission for Technical Co-operation in Africa South of the Sahara) held an important Conference on the subject in 1961 at Arusha in Tanganyika, attended by specialists from Africa and other continents, and by representatives, official and unofficial, of the majority of Sub-Saharan African territories. In this they had the strong support of F.A.O. (the Food and Agriculture Organisation of the United Nations), which is deeply concerned with the general shortage of protein in Africa, and the possibility of remedying the deficiency by proper management and game-cropping in wild life areas; and also of Unesco, which is now making conservation a major plank in its programme.

Here representatives of many countries had the opportunity of sharing their difficulties, pooling their experiences and discussing them with scientific experts and international officials.

Meanwhile, it is urgent to bring home the importance and multiple values of African wild life and wild nature to people in Africa, both Europeans and Africans, but especially Africans, including simple tribesmen as well as chiefs and elders, politicians and Ministers as well as office-workers and farmers.

This is an educational job. It can be done in various ways. In this long term it will be important to pay more attention to conservation in the educational curriculum at all levels; and all institutions of university level should provide institutes of African Studies, including the study of African nature and its conservation.

Technical training for those professionally concerned with conservation, and general courses on conservation and land-use for administrators and others whose work impinges on human ecology, are also essential. A beginning has recently been made with courses of the former type at the University College of Rhodesia and Nyasaland in Salisbury, and with those of the latter type at University College, London.

In the short term, much can be done by making films for Africans on wild life and its value, using mobile film units to show them to schools, African District Councils, meetings of chiefs, and other gatherings. Film-strips and posters are also very useful, and radio talks command an increasing audience.

But first-hand experience is best of all: and as many Africans as possible should be encouraged and enabled to see their own national parks. Visits of this sort have been arranged in Uganda, for college students and town workers, chiefs and schoolboys, and the results are encouraging. Most of the visitors have never seen any large wild animals, and are thrilled when they first do so. "Those are elephants— how wonderful! And they are *our* elephants. We must pro- tect them and help other Africans to enjoy the sight of them"—that, I am told, is the usual sort of reaction. But again, such visits cost money: and a lot more money is needed to make the flow of African visitors at all adequate.

The most practical method of convincing Africans of the value of wild life is to arrange for them to benefit financially or physiologically from its utilization, a matter which I dis- cussed earlier.

Another practical method is to set up local parks, the revenue and prestige from which would go to the local District Council authority. This has now been done in two splendid wildlife areas in Kenya Masailand, Amboseli and the Maru River Triangle. It has also been done at Kinna, in the Meru area of Kenya, where we were shown round by the European warden recently appointed by the African District Council. (It was here that Elsa, the celebrated lioness of Joy Adamson's *Born Free*, was liberated and produced her three cubs, sired by a wild lion.) The ideal would be for such areas to be run for the benefit of the local population, while retain- ing the international prestige of national parks. Attempts

195

should be made to achieve such a dual status, at the same time local and international, for the Gorilla Reserve in the Kigezi district of western Uganda, and for the incomparable Ngorongoro Crater. (Ngorongoro, by the way, is the second largest crater in the world, over eleven miles in diameter: I estimated that the wildebeest and zebra which dot its floor bear the same ratio to its total size as fleas would do to the dimensions of a large blue whale.)

Incomparable—that is the word for the African national parks and reserves in general. They are indeed a treasure for the world. Looking back on our two journeys in eastern Africa in 1929 and 1960, I think of the wonderful enrichment of experience they gave to us, and can give to millions of others. There was the Murchison Falls Park, with the mighty Nile compressed into a fall a mere twenty feet wide, with elephants raiding the hotel's dustbins, and launches taking you quietly into the closest proximity to hippos and crocodiles, fish-eagles and buffaloes.

There was Manyara in Tanganyika, with a comfortable hotel on the scarp of the western rift, looking down on elephants in a sector of virgin rain-forest, traversed by clear streams; beyond the forest a strip of plain with zebras, buffaloes and antelopes; and beyond that again the marshy lake margin with an inconceivable variety of water-birds.

At Amboseli there were wildebeest and giraffes just beyond the Warden's garden fence, rhinos and leopards and bat-eared foxes and marabous farther afield, Masai herdsmen with their cattle, and the immense dome of Kilimanjaro towering serenely into the blue. On Lake Nakuru there were half a million pink flamingoes, set in lovely scenery. Across the Congo border there were the seven great Virunga volcanoes striding across the floor of the western rift, with gorillas on their forested slopes and giant lobelias and senecios on their open mountain heathlands.

In the Queen Elizabeth Park there was a three-fold treasure—the Kazinga Channel with its elephants and hippos and water-birds, the uplands with their scores of small and medium green craters, haunts of elephants and many other wild creatures, and the flat southern plains with breathtaking herds of antelopes. In the Nairobi Park, a bare ten miles from the centre of the bustling city, there were lions and ostriches, cheetahs and zebras, impalas and giraffes, baboons and gazelles.

But why go on? Each park has some special glory to offer, and together they add up to give you and me, and anyone who takes the trouble to visit them, a priceless store of enriched experience and a unique glimpse of the world of nature as it was before the coming of man. They must at all costs be preserved.

TOYNBEE AND TIME-SCALES

READING Arnold Toynbee on civilization the other day,
I came on a passage which made me rub my eyes. His
thesis, in brief, is that all civilizations can be treated as contemporary, because it is only a few thousand years since
civilization first began, and that is infinitesimal in comparison
with the time-scale, involving hundreds of millions of years
of biological evolution, since life first began. "On this true
time-scale," he writes, "the events of 'ancient history' are
virtually contemporary with our own life-time."

This implies two things—first that the biological or
evolutionary time-scale is in some essential way true, while
our customary human time-scale is not true; and secondly
that there really isn't any historical time-scale at all, but only
a "virtual contemporaneity" of all historical events.

It so happened that I had just come back from the Middle
East, and that there I had been getting more and more
interested in the development of successive civilizations—
Sumerian, Egyptian, Assyrian, Syrian, Iranian, Cretan,
Greek. How could the historical facts be interpreted except
as part of a process in time, a dialectic sequence of events?
And must not a process in time have its appropriate time-
scale? I had also for some years been trying to come to some
conclusion about the relation of human history to biological
evolution, and I think Professor Toynbee is radically mis-
taken. Scientific reality *is*, under one essential aspect, a
process in time: we may call it evolution in the most general
sense of the word. Human history and biological evolution
are both segments of this single process of evolution: and so
is the inorganic evolution of nebulae and stars.

But they each have their own time-scale. The time-scale of
stellar evolution is about 10,000 times as extensive as that
of the evolution of life, and this in its turn about 100,000
times as extensive as that of human civilization. Major
changes in the development of the sun are measurable in

198

billions of years, those in the evolution of animals in hundreds of millions, those in the growth of civilization in millennia or hundreds of centuries.

But this does not mean that one time-scale is "truer" than the others—it means merely that the methods of evolution are different in the three segments, and result in different speeds of evolutionary change. The difference in the time-scales is thus a matter of operative convenience. It would be as impractical to try to discuss the age of the sun in terms of centuries as it would be to insist on specifying a man's age in seconds. It would be even more convenient to have a single scientific terminology in which to express these quantitative differences in time-scales, just as we have a single scientific terminology—the metric—to express quantitative differences in size. As I have suggested elsewhere, we could use the term *cron*, meaning a million years, as the basic unit of evolutionary time, with kilochron for a thousand million (American billion) years, millicron for a millennium, and microchron for a year.

The new method available to life, as against lifeless matter, was that of natural selection, which speeded up change enormously by permitting the accumulation of separate favourable mutations in the single evolving stream of living matter that we call the species or the race.

J. B. S. Haldane has pointed out that we can now begin to calculate the actual rates of change in those lines of biological evolution where fossil remains are abundant: we can, for example, determine the percentage increase in length or breadth of a bone over so many million years, as measured by data on radioactivity in the rocks. In ancestral horses, for instance, such increases all fall between 1 and 10 per cent. per million years. (Haldane suggests calling the unit of evolutionary rate a *darwin*, in which case the average for the horse stock would be around 40 millidarwins.)

The evolution of civilization (as measured by social, cultural and technological changes) in its turn proceeds on a new and much abbreviated time-scale. The method which permits it to do this is of course the social transmission of tradition, in the broad sense of the word. This can be cumulative in its effects, which means that evolutionary change is accelerated, instead of remaining approximately constant in rate over a whole segment of evolution, as was the case in biological evolution. The transition from biological to human evolution began when man became truly man, through becoming capable of speech and conceptual thought, and therefore capable of transmitting tradition. At first, however, the rate of change, though far higher than in biological evolution, was slow by modern standards. During the entire Lower Paleolithic, a period of something like half a million years, technological advance, as measured by the improvement of stone implements, must have been quite imperceptible from one generation to the next. Even so, the time-unit for effective improvement was of the order of 100,000 years instead of about ten million years as with biological evolution.

With the advent of modern man in the Upper Paleolithic, the rate of change was speeded up some tenfold, and perhaps another tenfold during the ten thousand years or so after the end of the last glacial period.

But it was the development of civilization—settled life, with writing and permanent buildings—which gave the new method of evolution by tradition its real chance. The time-scale of human change became measurable in centuries, and eventually, with the advent of printing and modern science, in decades, until today the rate of change is bewilderingly rapid.

But this does not prevent human civilization from having its time-scale just as much as biological evolution. Indeed, since civilization is a process, moving irreversibly through time, it must have its time-scale, and therefore its tempo. Nothing is gained by pretending that later developments of civilization are really, or virtually, or in any way whatsoever, contemporaneous with earlier ones. What does, of course, occur, as I have pointed out elsewhere in this volume, is that an earlier type or grade of civilization or cultural organization may persist for long periods and so become secondarily contemporaneous with a later type: however, in their historical origins the two types were not contemporaneous, but successive.

On the other hand, a great deal may be gained if, for instance, we try to find out what is the most favourable tempo for the evolution of civilization, what is the best balance to strike between stability and change. In biological evolution, such a balance appears to have been automatically struck by natural selection adjusting the mutation-rate. In our own evolution, we shall have consciously to plan and execute the method for controlling the release of change.

Perhaps we shall never be successful; but anyhow we shall have a very interesting time studying the time-scales of different human activities at different periods—of art as against science, for instance, in ancient Egypt, in classical Athens, in the Renaissance, in the Western world, and in the U.S.S.R. today. The scientific investigation of historical time-scales will, I feel sure, be a main concern of future Toynbees.

TEILHARD DE CHARDIN

The Phenomenon of Man is a remarkable work by a remarkable human being. Père Teilhard de Chardin was at the same time a Jesuit Father and a distinguished palaeontologist. In *The Phenomenon of Man* he has aimed at a threefold synthesis—of the material and physical world with the world of mind and spirit; of the past with the future; and of variety with unity, the many with the one. He achieves this by examining every fact and every subject of his investigation *sub specie evolutionis*, with reference to its development in time and to its evolutionary position. Conversely, he is able to envisage the whole of knowable reality not as a static mechanism but as a process. In consequence, he is driven to search for human significance in relation to the trends of that enduring and comprehensive process; the measure of his stature is that he has so largely succeeded in the search.

Père Teilhard starts from the position that mankind in its totality is a phenomenon to be described and analysed like any other phenomenon: it and all its manifestations, including human history and human values, are proper objects for scientific study.

His second and perhaps most fundamental point is the absolute necessity of adopting an evolutionary point of view. Though for certain limited purposes it may be useful to think of phenomena as isolated statically in time, they are in point of fact never static: they are always processes or parts of processes. The different branches of science combine to demonstrate that the universe in its entirety must be regarded as one gigantic process, a process of becoming, of attaining new levels of existence and organization, which can properly be called a genesis or an evolution. For this reason, he uses words like *noogenesis*, to mean the gradual evolution of mind or mental properties, and repeatedly stresses that we should no longer speak of a cosmology but of a *cosmogenesis*. Similarly, he likes to use a pregnant term like *hominization* to denote 'the process by which the original proto-human stock became (and is still becoming) more truly human, the process by which potential man realized more and more of his possibilities. Indeed, he extends this evolutionary terminology by employing terms like *ultra-hominization* to denote the deducible future stage of the process in which man will have so far transcended himself as to demand some new appellation.

With this approach he is rightly and indeed inevitably driven to the conclusion that, since evolutionary phenomena (of course including the phenomenon known as man) are processes, they can never be evaluated or even adequately described solely or mainly in terms of their origins: they must be defined by their direction, their inherent possibilities (including of course also their limitations), and their deducible future trends. He quotes with approval Nietzsche's view that man is unfinished and must be surpassed or completed; and proceeds to deduce the steps needed for this completion.

Père Teilhard was keenly aware of the importance of vivid and arresting terminology. Thus in 1925 he independently coined the term *noosphere* to denote the sphere of mind, as opposed to, or rather superposed on, the biosphere or sphere of life, and acting as a transforming agency promoting hominization (or, as I would put it, progressive psychosocial evolution). He may perhaps be criticized for not defining the term more explicitly. By *noosphere* did he intend simply the total pattern of thinking organisms (i.e. human beings) and their activity, including the patterns of their inter-relations: or did he intend the special environment of man, the systems of organized thought and its products in which men move and have their being, as fish swim and reproduce in rivers and the sea?[1] Perhaps it might have been better to restrict

---

[1] In *Le Phénomène Humain* (p. 201) he refers to the *noosphere* as a new layer or membrane on the earth's surface, a "thinking layer" superposed on the living layer of the *biosphere* and the lifeless layer of inorganic material, the *lithosphere*. But in his earlier formulation of 1925, in *La Vision du Passé* (p. 92), he calls it "une sphère de la réflexion, de l'invention consciente, de l'union sentie des âmes".

noosphere to the first-named sense, and to use something like *noosystem* for the second. But certainly *noosphere* is a valuable and thought-provoking word.

He usually uses *convergence* to denote the tendency of mankind, during its evolution, to superpose centripetal on centrifugal trends, so as to prevent centrifugal differentiation from leading to fragmentation, and eventually to incorporate the results of differentiation in an organized and unified pattern. Human convergence was first manifested on the genetic or biological level: after *Homo sapiens* began to differentiate into distinct races (or *subspecies*, in more scientific terminology) migration and intermarriage prevented the pioneers from going further, and led to increasing interbreeding between all human variants. As a result, man is the only successful type which has remained as a single interbreeding group or species, and has not radiated out into a number of biologically separated assemblages (like the bird type with about 8600 species, or the insect type with over half a million).

Cultural differentiation set in later, producing a number of psychosocial units with different cultures. However, these "inter-thinking groups", as one writer has called them, are never so sharply separated as are biological species; and with time, the process known to anthropologists as cultural diffusion, facilitated by migration and improved communications, led to an accelerating counter-process of cultural convergence, and so towards the union of the whole human species into a single inter-thinking group based on a single self-developing framework of thought (or noosystem).

In parenthesis, Père Teilhard showed himself aware of the danger that this tendency might destroy the valuable results of cultural diversification, and lead to drab uniformity instead of to a rich and potent pattern of variety-in-unity. However, perhaps because he was so deeply concerned with establishing a global unification of human awareness as a necessary prerequisite for any real future progress of mankind, and perhaps also because he was by nature and inclination more interested in rational and scientific thought than in the arts, he did not discuss the evolutionary value of cultural variety in any detail, but contented himself by maintaining that East and West are culturally complementary, and that both are needed for the further synthesis and unification of world thought.

Before passing to the full implications of human convergence, I must deal with Père Teilhard's valuable but rather difficult concept of *complexification*. This concept includes, as I understand it, the genesis of increasingly elaborate organization during cosmogenesis, as manifested in the passage from subatomic units to atoms, from atoms to inorganic and later to organic molecules, thence to the first subcellular living units or self-replicating assemblages of molecules, and then to cells, to multicellular individuals, to cephalized metazoa with brains and incipient minds, to primitive man, and now to civilized societies.

But it involves something more. He speaks of complexification as an all-pervading tendency, involving the universe in all its parts in an *enroulement organique sur soi-même*, or by an alternative metaphor, as a *reploiement sur soi-même*. He thus envisages the world-stuff as being "rolled up" or "folded in" upon itself, both locally and in its entirety, and adds that the process is accompanied by an increase of energetic "tension" in the resultant "corpuscular" organizations, or individualized constructions of increased organizational complexity. For want of a better English phrase, I shall use *convergent integration* to define the operation of this process of self-complexification.

Père Teilhard also maintains that complexification by convergent integration leads to the intensification of mental subjective activity—in other words to the evolution of progressively more conscious mind. Thus he states that full consciousness (as seen in man) is to be defined as "the specific effect of organized complexity". But, he continues, comparative study makes it clear that higher animals have minds of a sort, and evolutionary fact and logic demand that minds should have evolved gradually as well as bodies and that accordingly mind-like (or "mentoid", to employ a barbarous word that I am driven to coin because of its usefulness) properties must be present throughout the universe. Accordingly we must envisage the intensification of mind, the raising of mental potential, as being the necessary consequence of complexification, operating by the convergent integration of increasingly complex units of organization.

The sweep of his thought goes even further. He seeks to link the evolution of mind with the concept of energy. If I understand him aright, he envisages two forms of energy, or perhaps two modes in which it is manifested—energy in the physicists' sense, measurable or calculable by physical methods, and "psychic energy" which increases with the complexity of organized units.[1] This view admittedly involves speculation of great intellectual boldness, but the speculation is extrapolated from a massive array of fact, and is disciplined by logic. It is, if you like, visionary: but it is the product of a comprehensive and coherent vision.

It might have been better to say that complexity of a sort is a necessary prerequisite for mental evolution rather than its cause. Some biologists, indeed, would claim that mind is generated solely by the complexification of certain types of organization, namely brains. However, such logic appears to me narrow. The brain alone is not responsible for mind, even though it is a necessary organ for its manifestation. Indeed an isolated brain is a piece of biological nonsense, as meaningless as an isolated human individual. As I have explained in another essay in this volume, I would prefer to say that mind is generated by or in complex organizations of living matter, capable of receiving information of many qualities or modalities about events both in the outer world and in itself, of synthesizing and processing that information in various organized forms, and of utilizing it to direct present and future action—in other words, by higher animals with their sense-organs, brains, glands, and muscles. Perhaps, indeed, organizations of such complexity can only arise in evolution when their construction enables them to incorporate and interiorize varied external information: certainly no non-living, non-sentient organization has reached anything like this degree of elaboration.

[1] See, e.g., C. Cuénot, *Pierre Teilhard de Chardin*, Paris, 1958, p. 430. We certainly need some new terms in this field: perhaps *neurergy* and *psychergy* would serve.

In human or psychosocial evolution, convergence has certainly led to increased complexity. In Père Teilhard's view, the increase of human numbers combined with the improvement of human communications has fused all the parts of the noosphere together, has increased the tension within it, and has caused it to become "infolded" upon itself, and therefore more highly organized. In the process of convergence and coalescence, what we may metaphorically describe as the psychosocial temperature rises. Mankind as a whole will accordingly achieve more intense, more complex, and more integrated mental activity, which can guide the human species up the path of progress to higher levels of hominization.

Père Teilhard was a strong visualizer. He saw with his mind's eye that "the banal fact of the earth's roundness"— the sphericity of man's environment—was bound to cause this intensification of psychosocial activity. In an unlimited environment, man's thought and his resultant psychosocial activity would simply diffuse outwards: it would extend over a greater area, but would remain thinly spread. But when it is confined to spreading out over the surface of a sphere, idea will encounter idea, and the result will be an organized web of thought, a noetic system operating under high tension, a piece of evolutionary machinery capable of generating high psychosocial energy. When I read his discussion of the subject, I visualized this selective web of living thought as the bounding structure of evolving man, marking him off from the rest of the universe and yet facilitating exchange with it: playing the same sort of role in delimiting the human unit of evolution and yet encouraging the complexification of its contents, as does the cell-membrane for the animal cell.

Years later, when at the University of California in 1952, this same vivid imagination led Père Teilhard to draw a parallel between the cyclotron generating immense intensities of physical energy in the inwardly accelerating spiral orbits of its fields of force, and the entire noosphere with its fields of thought curved round upon themselves to generate new levels of "psychical energy". How his imagination would have kindled at the sight of the circular torus of Zeta, within whose bounding curves are generated the highest physical energies ever produced by man!

Père Teilhard, extrapolating from the past into the future, envisaged the process of human convergence as tending to a final state[1] which he called *"point Omega"*, as opposed to the *Alpha* of elementary material particles and their energies. If I understand him aright, he considers that two factors are co-operating to promote this further complexification of the noosphere. One is the increase of knowledge about the universe at large, from the galaxies and stars to human societies and individuals. The other is the increase of psycho-social pressure on the surface of our planet. The result of the one is that the noosphere incorporates ever more facts of the cosmos, including the facts of its general direction and its trends in time, so as to become more truly a microcosm, which (like all incorporated knowledge) is both a mirror and a directive agency. The result of the other is the increased unification and the increased intensity of the system of human thought. The combined result, according to Père Teilhard, will be the attainment of point Omega, where the noosphere will be intensely unified and will have achieved a "hyperpersonal" organization.

Here his thought is by no means clear to me. Sometimes he seems to equate this future hyperpersonal psychosocial organization with an emergent Divinity: at one place, for instance, he speaks of the trend as a Christogenesis; and elsewhere he appears not to be guarding himself sufficiently against the dangers of personifying the nonpersonal elements of reality. Sometimes, too, he seems to envisage as desirable the merging of individual human variety in this new unity. Many scientists may, as I do, find it impossible to follow him all the way in his attempt to reconcile the supernatural elements in Christianity with the facts and implications of

[1] Presumably, in designating this state as Omega, he believed that it was a truly final condition. It might have been better to think of it merely as a novel state or mode of organization, beyond which the human imagination cannot at present pierce, though perhaps the strange facts of extra-sensory perception unearthed by the infant science of parapsychology may give us a clue as to a possible more ultimate state.

evolution; but this does not detract from the positive value of his naturalistic general approach.

This concept of a hyperpersonal mode of organization sprang from Père Teilhard's conviction of the supreme importance of personality. A developed human being, as he rightly pointed out, is not merely a more highly individualized individual. He has crossed the threshold of self-consciousness to a new mode of thought, and as a result has achieved some degree of conscious integration—integration of the self with the outer world of men and nature, integration of the separate elements of the self with each other. He is a person, an organism which has transcended individuality in personality. This attainment of personality was an essential element in man's past and present evolutionary success: accordingly its fuller achievement must be an essential aim for his evolutionary future.

This belief in the pre-eminent importance of the personality in the scheme of things was for him a matter of faith, but of faith supported by rational enquiry and scientific knowledge. It prevented him from diluting his concept of the divine principle inherent in reality, in a vague and meaningless pantheism, just as his apprehension of the entire process of reality as a system of interrelations, and of mankind as actively participating in that process, saved him from losing his way in the deserts of individualism and existentialism.

He realized that the appearance of human personality was the culmination of two major evolutionary trends—the trend towards more extreme individuation, and that towards more extensive interrelation and co-operation: persons are individuals who transcend their merely organic individuality in conscious participation.

His understanding of the method by which organisms become first individualized and then personalized gave him a number of valuable insights. Basically, the process depends on cephalization—the differentiation of a head as the dominant guiding region of the body, forwardly directed, and containing the main sense-organs providing information about the outer world and also the main organ of co-ordination or brain.

With his genius for fruitful analogy, he points out that the process of evolution on earth is itself now in the process of becoming cephalized. Before the appearance of man, life consisted of a vast array of separate branches, linked only by an unorganized pattern of ecological interaction. The incipient development of mankind into a single psychosocial unit, with a single noosystem or common pool of thought, is providing the evolutionary process with the rudiments of a head. It remains for our descendants to organize this global noosystem more adequately, so as to enable mankind to understand the process of evolution on earth more fully and to direct it more adequately.

I had independently expressed something of the same sort, by saying that in modern scientific man, evolution was at last becoming conscious of itself—a phrase which I found delighted Père Teilhard. His formulation, however, is more profound and more seminal: it implies that we should consider inter-thinking humanity as a new type of organism, or rather a new type of living organization, whose destiny it is to realize new possibilities for evolving life on this planet. Accordingly, we should endeavour to equip it with the mechanisms necessary for the proper fulfilment of its task—the psychosocial equivalents of sense-organs, emotive organs, effector organs, and a central nervous system with dominant brain; and our aim should be the gradual personalization of the human evolutionary type—its conversion, on the new level of co-operative inter-thinking, into the equivalent of a person.

Once he had grasped and faced the fact of man as an evolutionary phenomenon, the way was open towards a new and comprehensive system of thought. It remained to draw the fullest conclusions from this central concept of man as the spearhead of evolution on earth, and to follow out the implications of this approach in as many fields as possible. The biologist may consider that in *The Phenomenon of Man* he paid insufficient attention to genetics and the possibilities and limitations of natural selection,[1] the theologian that his treatment of the problems of sin and suffering was inadequate

[1] Though in his Institute for Human Studies he envisaged a section of Eugenics.

or at least unorthodox, the social scientist that he failed to take sufficient account of the facts of political and social history. But he saw that what was needed at the moment was a broad sweep and a comprehensive treatment. This was what he essayed in *The Phenomenon of Man*. In my view he achieved a remarkable success, and opened up vast territories of thought to further exploration and detailed mapping.

*　　*　　*

The facts of Père Teilhard's life help to illuminate the development of his thought. His father was a small land-owner in Auvergne, a gentleman farmer who was also an archivist, with a taste for natural history. Pierre was born in 1881, the fourth in a family of eleven. At the age of 10 he went as a boarder to a Jesuit College where, besides doing well in all prescribed subjects of study, he became devoted to field geology and mineralogy. When 18 years old, he decided to become a Jesuit, and entered their order. At the age of 24, after an interlude in Jersey mainly studying philosophy, he was sent to teach physics and chemistry in a Jesuit College at Cairo. In the course of his three years in Egypt, and a further four studying theology in Sussex, he acquired real competence in geology and palaeontology; and before being ordained priest in 1912, a reading of Bergson's *Evolution Créatrice* had helped to inspire in him a profound interest in the general facts and theories of evolution. Returning to Paris, he pursued his geological studies and started working under Marcellin Boule, the leading prehistorian and archaeologist of France, in his Institute of Human Palaeontology at the Museum of Natural History. It was here that he met his life-long friend and colleague in the study of pre-history, the Abbé Breuil, and that his interests were first directed to the subject on which his life's work was centred —the evolution of man. In 1913 he visited the site where the famous (and now notorious) Piltdown skull had recently been unearthed, in company with its discoverer Dr Dawson and the leading English palaeontologist Sir Arthur Smith Woodward. This was his first introduction to the excitements of palaeontological discovery and scientific controversy.

During the First World War he served as a stretcher-bearer, receiving the Military Medal and the Legion of Honour, and learnt a great deal about his fellow-men and about his own nature. The war strengthened his sense of religious vocation, and in 1918 he made the triple vow of poverty, chastity and obedience.

By 1919 the major goals of his life were clearly indicated. Professionally, he had decided to embark on a geological career, with special emphasis on palaeontology. As a thinker, he had reached a point where the entire phenomenal universe, including man, was revealed as a process of evolution, and he found himself impelled to build up a generalized theory or philosophy of evolutionary process which would take account of human history and human personality as well as of biology, and from which one could draw conclusions as to the future evolution of man on earth. And as a dedicated Christian priest, he felt it imperative to try to reconcile Christian theology with this evolutionary philosophy, to relate the facts of religious experience to those of natural science.

Returning to the Sorbonne, he took his Doctorate in 1922. He had already become Professor of Geology at the Catholic Institute of Paris, where his lectures attracted great attention among the students. In 1923, however, he went to China for a year on behalf of the Museum, on a palaeontological mission directed by another Jesuit, Père Licent. His *Lettres de Voyage* reveal the impression made on him by the voyage through the tropics, and by his first experience of geological research in the desert remoteness of Mongolia and north-western China. This expedition inspired *La Messe sur le Monde*, a remarkable and truly poetical essay which was at one and the same time mystical and realistic, religious and philosophical.

A shock awaited him after his return to France. Some of the ideas which he had expressed in his lectures, about original sin and its relation to evolution, were regarded as unorthodox by his religious superiors, and he was forbidden to continue teaching. In 1926 he returned to work with Père Licent in China, where he was destined to stay, with brief returns to France and excursions to the United States, to

Abyssinia, India, Burma and Java, for twenty years. Here, as scientific adviser to the Geological Survey of China, centred first at Tientsin and later at Peking, he met and worked with outstanding palaeontologists of many nations, and took part in a number of expeditions, including the Citroën *Croisière Jaune* under Haardt, and Davidson Black's expedition which unearthed the skull of Peking man.

In 1938 he was appointed Director of the Laboratory of Advanced Studies in Geology and Palaeontology in Paris, but the outbreak of war prevented his return to France. His enforced isolation in China during the six war years, painful and depressing though it often was, undoubtedly helped his inner spiritual development (as the isolation of imprisonment helped to mature the thought and character of Nehru and many other Indians). It encouraged ample reading and reflection, and stimulated the full elaboration of his thought.

It was a nice stroke of irony that the action of Père Teilhard's religious superiors in barring him from teaching in France because of his ideas on human evolution should have led him to China, brought him into intimate association with one of the most important discoveries in that field, and driven him to enlarge and consolidate his "dangerous thoughts".

During the whole of this period he was writing essays and books on various aspects and implications of evolution, culminating in 1938 in the manuscript of *Le Phénomène Humain*. But he never succeeded in obtaining permission to publish any of his controversial or major works. This caused him much distress, for he was conscious of a prophetic mission: but he faithfully observed his vow of obedience. Professionally too he was extremely active throughout this period. He contributed a great deal to our knowledge of palaeolithic cultures in China and neighbouring areas, and to the general understanding of the geology of the Far East. This preoccupation with large-scale geology led him to take an interest in the geological development of the world's continents: each continent, he considered, had made its own special contribution to biological evolution. He also did important palaeontological work on the evolution of various mammalian groups.

The wide range of his vision made him impatient of over-specialization, and of the timidity which refuses to pass from detailed study to broad synthesis. With his conception of mankind as at the same time an unfinished product of past evolution and an agency of distinctive evolution to come, he was particularly impatient of what he felt as the narrowness of those anthropologists who limited themselves to a study of physical structure and the details of primitive social life. He wanted to deal with the entire human phenomenon, as a transcendence of biological by psychosocial evolution. And he had considerable success in redirecting along these lines the institutions with which he was connected.

Back in France in 1946, Père Teilhard plunged eagerly into European intellectual life, but in 1947 he had a serious heart attack, and was compelled to spend several months convalescing in the country. On his return to Paris, he was enjoined by his superiors not to write any more on philo-sophical subjects: and in 1948 he was forbidden to put forward his candidature for a Professorship in the College de France in succession to the Abbé Breuil, though it was known that this, the highest academic position to which he could aspire, was open to him. But perhaps the heaviest blow awaited him in 1950, when his application for permission to publish *Le Groupe Zoologique Humain* (a recasting of *Le Phénomène Humain*) was refused in Rome. By way of com-pensation he was awarded the signal honour of being elected Membre de l'Institut, as well as having previously become a Corresponding Member of the Académie des Sciences, an officer of the Légion d'Honneur, and a director of research in the Centre National de la Recherche Scientifique.

Already in 1948 he had been invited to visit the U.S.A., where he made his first contacts with the Wenner-Gren Foundation (or Viking Foundation as it was then called), in whose friendly shelter he spent the last four years of his life. The Wenner-Gren Foundation also sponsored his two visits to South Africa, where he was able to study at first hand the remarkable discoveries of Broom and Dart concerning Australopithecus, that near-ancestor of man, and to lay down a plan for the future co-ordination of palaeontological and archaeological work in this area, so important as a centre of hominid evolution.

His position in France became increasingly difficult, and in 1951 he moved his headquarters to New York. Here, at the Wenner-Gren Foundation, he played an important role in framing anthropological policy, and made valuable contributions to the international symposia which it organized. And here, in 1954, I had the privilege of working with him in one of the remarkable discussion groups set up as part of the Columbia Bicentennial celebrations. Just before this, he had returned to France for a brief but stimulating month of discussion.

Throughout this period, he had been actively developing his ideas, and had written his spiritual autobiography, *Le Coeur de la Matière*, the semi-technical *Le Groupe Zoologique Humain*, and various technical and general articles later included in the collections entitled *La Vision du Passé* and *L'Apparition de l'Homme*.

He was prevailed on to leave his manuscripts to a friend. They therefore could be published after his death, since permission to publish is only required for the work of a living writer. The prospect of eventual publication must have been a great solace to him, for he certainly regarded his general and philosophical writings as the keystone of his life's work, and felt it his supreme duty to proclaim the fruits of his labour.

*   *   *

Père Teilhard's influence on the world's thinking is bound to be important. Through his combination of wide scientific knowledge with deep religious feeling and a rigorous sense of values, he has forced theologians to view their ideas in the new perspective of evolution, and scientists to see the spiritual implications of their knowledge. He has helped both to clarify and to unify our vision of reality. In the light of that new comprehension, it is no longer possible to maintain that science and religion must operate in thought-tight compartments or concern separate sectors of life; they are both relevant to the whole of human existence. The religiously-

215

minded can no longer turn their backs upon the natural world, or seek escape from its imperfections in a supernatural world; nor can the materialistically-minded deny importance to spiritual experience and religious feeling.

Like him, we must face the phenomena. If we face them resolutely, and avail ourselves of the help which the intellectual and spiritual travail of men like Père Teilhard has provided, we shall find a more assured basis for our thought and a more certain direction for our evolutionary advance. But, like him, we must not take refuge in abstractions or generalities. He always took account of the specific realities of man's present situation, though set against the more general realities of long-term evolution; and he always endeavoured to think concretely, in terms of actual patterns of organization—their development, their mode of operation, and their effects.

As a result, he has helped us to define more adequately both our own nature, the general evolutionary process, and our place and role in it. Thus clarified, the evolution of life becomes a comprehensible phenomenon. It is an anti-entropic process, running counter to the second law of thermodynamics with its degradation of energy and its tendency to uniformity. With the aid of the sun's energy, biological evolution marches uphill, producing increased variety and higher degrees of organization.

It also produces more varied, more intense, and more highly organized mental activity or awareness. During evolution, awareness (or if you prefer, the mental properties of living matter) becomes increasingly important to organisms, until in mankind it becomes the most important characteristic of life, and gives the human type its dominant position.

After this critical point has been passed, evolution takes on a new character: it becomes primarily a psychosocial process, based on the cumulative transmission of experience and its results, and working through an organized system of awareness, a combined operation of knowing, feeling, and willing.

On this new psychosocial level, the evolutionary process

leads to new types and higher degrees of organization. On the one hand there are new patterns of co-operation among individuals—co-operation for practical control, for enjoyment, for education, and notably in the last few centuries, for obtaining new knowledge; and on the other there are new patterns of thought, new organizations of awareness and its products.

As a result, new and often wholly unexpected possibilities have been realized, the variety and degree of human fulfilment has been increased. Père Teilhard helps us to see which possibilities are in the long run desirable. What is more, he has helped to define the conditions of advance, the conditions which will permit an increase of fulfilment and prevent an increase of frustration. The conditions of advance are these: global unity of mankind's noetic organization or system of awareness, but a high degree of variety within that unity; love, with goodwill and full co-operation; personal integration and internal harmony; and increasing knowledge.

Knowledge is basic. It is knowledge which enables us to understand the world and ourselves, and to exercise some control or guidance. It sets us in a fruitful and significant relation with the enduring processes of the universe. And, by revealing the possibilities of fulfilment that are still open, it provides an overriding incentive.

We, mankind, contain the possibilities of the earth's immense future, and can realize more and more of them on condition that we increase our knowledge and our love. That, it seems to me, is the distillation of *The Phenomenon of Man*.

THE NEW DIVINITY

THE Bishop of Woolwich's courageous book, *Honest to God*, is impressive evidence not merely of what he calls our present theological ferment, but of the general ideological ferment and indeed of the revolution of thought through which we are struggling.

This is the inevitable outcome of the new vision of the world and man's place and role in that world—in a word, of man's destiny—which our new knowledge has revealed. This new vision is both comprehensive and unitary. It integrates the fantastic diversity of the world into a single framework, the pattern of all-embracing evolutionary process. In this unitary vision, all kinds of splits and dualisms are healed. The entire cosmos is made out of one and the same world-stuff, operated by the same energy as we ourselves. "Mind" and "matter" appear as two aspects of our unitary mind-bodies. There is no separate supernatural realm: all phenomena are part of one natural process of evolution. There is no basic cleavage between science and religion; they are both organs of evolving humanity.

This earth is one of the rare spots in the cosmos where mind has flowered. Man is a product of nearly three billion years of evolution, in whose person the evolutionary process has at last become conscious of itself and its possibilities. Whether he likes it or not, he is responsible for the whole further evolution of our planet.

*   *   *

Dr Robinson describes the current image of God as follows: "Somewhere beyond this universe is a Being, a centre of personal will and purpose, who created it and sustains it, who loves it and who 'visited' it in Jesus Christ. But I need not go on, for this is 'our' God. Theism means being convinced that this Being exists: atheism means denying that he does." However, he continues as follows: "But I suspect that we have reached the point where this mental image of God is also more of a hindrance than a help. . . . Any image can become an idol, and I believe that Christians must go through the agonizing process in this generation of detaching themselves from this idol." He even writes that he heartily agrees with something I wrote many years ago in my *Religion Without Revelation*—"The sense of spiritual relief which comes from rejecting the idea of God as a superhuman being is enormous."

And yet he clings to the essentially personal concept of God—"nothing," he writes, "can separate us from the love of God"; and sums up his position in the following assertion, that "God is ultimate reality . . . and ultimate reality must exist".

To the implications of these statements I shall return. Meanwhile let me state the position as I see it. Man emerged as the dominant type on earth about a million years ago, but has only been really effective as a psychosocial organism for under ten thousand years. In that mere second of cosmic time, he has produced astonishing achievements—but has also been guilty of unprecedented horrors and follies. And looked at in the long perspective of evolution he is singularly imperfect, still incapable of carrying out his planetary responsibilities in a satisfactory manner.

The radical evolutionary crisis through which man is now passing can only be surmounted by an equally radical reorganization of his dominant system of thought and belief. During human history, there has been a succession of dominant systems of thought and belief, each accompanying a new organization of social, political and economic activities—agriculture with its rituals of rebirth as against hunting with its magic; early civilization with its cities and sacred kings, its written records and its priesthoods; universal and monotheist religion; later, the scientific, the industrial and the technological revolutions with their corresponding patterns of thought; and now the evolutionary and humanist revolution, whose ideological and social implications have still to be thought out.

What has all this to do with Dr Robinson's views on God,

or indeed with religion at all? The answer is, a great deal. In the first place, religion in some form is a universal function of man in society, the organ for dealing with the problems of destiny, the destiny of individual men and women, of societies and nations, and of the human species as a whole. Religions always have some intellectual or ideological framework, whether myth or theological doctrine; some morality or code of behaviour, whether barbaric or ethically rationalized; and some mode of ritualized or symbolic expression, in the form of ceremonial or celebration, collective devotion or thanksgiving, or religious art. But, as the history and comparative study of religions makes clear, the codified morality and the ritualized expression of a religion, and indeed in the long run its social and personal efficacy, derives from its "theological" framework. If the evolution of its ideological pattern does not keep pace with the growth of knowledge, with social change and the march of events, the religion will increasingly cease to satisfy the multitudes seeking assurance about their destiny, and will become progressively less effective as a social organ.

Eventually the old ideas will no longer serve, the old ideological framework can no longer be tinkered up to bear the weight of the facts, and a radical reconstruction becomes necessary, leading eventually to the emergence of a quite new organization of thought and belief, just as the emergence of new types of bodily organization was necessary to achieve biological advance.

Such major organizations of thought may be necessary in science as much as in religion. The classical example, of course, was the re-patterning of cosmological thought which demoted the earth from its central position and led to the replacement of the geocentric pattern of thought by a heliocentric one. I believe that an equally drastic reorganization of our pattern of religious thought is now becoming necessary, from a god-centred to an evolution-centred pattern. Simplified down to its bare essentials the stepwise reorganization of Western religious thought seems to have proceeded as follows. In its early, paleolithic stage religion was magic-centred, based on the ideas of magic force inherent in nature, in personages such as "medicine men" and shamans, and in human incantations, spells and other magic practices, including witchcraft. This type of belief developed gradually into animism and so to polydaimonism and polytheism; while with the coming of agriculture a new pattern was imposed, centering on the ideas of fertility and rebirth, and leading to the rise of priest-kings and eventually of divinized monarchs. The next major revolution of religious thought came in the first millennium B.C. with the independent rise of monotheist and/or universalist religions, culminating in Christianity and later branching off into Islam. The last two thousand years have seen the development of elaborate monotheistic theologies; but in the process their single God has broken into many, or at least has assumed a number of distinct and indeed sometimes actively hostile forms; and their nominal universalism has degenerated into competition for the possession of absolute truth.

Of course a great deal of magic survived into the polytheist priest-king stage, and some persists in thinly disguised form in Christian and Mohammedan practices and ideas today. Similarly, elements of polydaemonism and polytheism persist in the nominally monotheist religion of Christianity, in the doctrine of the Trinity (with the virtual divinization of the Virgin in Catholicism), in the multiplication of its Saints and Angels, and in so doing has increased its flexibility.

But to come back to Dr Robinson. He is surely right in concentrating on the problem of God, rather than on the resurrection or the after-life, for God is Christianity's central hypothesis.

But he is surely wrong in making such statements as that "God is ultimate reality". God is a hypothesis constructed by man to help him understand what existence is all about. The god hypothesis asserts the existence of some sort of supernatural personal or superpersonal being, exerting some kind of purposeful power over the universe and its destiny. To say that God is ultimate reality is just semantic cheating, as well as being so vague as to become effectively meaningless (and when Dr Robinson continues by saying "and ultimate reality must exist" he is surely running round a philosophically very vicious circle).

Dr Robinson, like Dr Tillich and many other modernist theologians, seems to me, and indeed to any humanist, to be trying to ride two horses at once, to keep his cake and eat it. He wants to be modern and meet the challenge of our new knowledge by stripping the image of God of virtually all its spatial, material, mythological, Freudian and anthropo-morphic aspects. But he still persists in retaining the term *God*, in spite of all its implications of supernatural power and personality; and it is these implications, not the modernists' fine-spun arguments, which consciously or unconsciously affect the ordinary man and woman. Heads I win, tails you lose: humanists dislike this elaborate double-talk. The ambiguity involved can be simply illustrated by substituting some of the modernists' definitions of God for the plain word itself. I am sure that many opponents of freer divorce use the phrase "whom God has joined together, let no man put asunder". If they were to proclaim that "whom universal reality has joined together, let no man put asunder", it would not carry the same weight.

Today the god hypothesis has ceased to be scientifically tenable, has lost its explanatory value and is becoming an intellectual and moral burden to our thought. It no longer convinces or comforts, and its abandonment often brings a deep sense of relief. Many people assert that this abandon-ment of the god hypothesis means the abandonment of all religion and all moral sanctions. This is simply not true. But it does mean, once our relief at jettisoning an outdated piece of ideological furniture is over, that we must construct some-thing to take its place.

Though gods and God in any meaningful sense seem destined to disappear, the stuff of divinity out of which they have grown and developed remains. This religious raw material consists of those aspects of nature and those experi-ences which are usually described as divine. Let me remind my readers that the term *divine* did not originally imply the existence of gods: on the contrary, gods were constructed to interpret man's experiences of this quality.

Some events and some phenomena of outer nature transcend ordinary explanation and ordinary experience. They inspire awe and seem mysterious, explicable only in terms of something beyond or above ordinary nature.

Such magical, mysterious, awe-inspiring, divinity-suggesting facts have included wholly outer phenomena like volcanic eruptions, thunder, and hurricanes; biological phenomena such as sex and birth, disease and death; and also inner, psychological phenomena such as intoxication, possession, speaking with tongues, inspiration, insanity, and mystic vision.

With the growth of knowledge most of these have ceased to be mysterious so far as rational or scientific explicability is concerned (though there remains the fundamental mystery of existence, notably the existence of mind). However, it is a fact that many phenomena are charged with some sort of magic or compulsive power, and do introduce us to a realm beyond our ordinary experience. Such events and such experiences merit a special designation. For want of a better, I use the term *divine*, though this quality of divinity is not truly supernatural but *transnatural*—it grows out of ordinary nature, but transcends it. The divine is what man finds worthy of adoration, that which compels his awe.

Much of every religion is aimed at the discovery and safeguarding of divinity in this sense, and seeks contact and communion with what is regarded as divine. A humanist evolution-centred religion too needs divinity, but divinity without God. It must strip the divine of the theistic qualities which man has anthropomorphically projected into it, search for its habitations in every aspect of existence, elicit it, and establish fruitful contact with its manifestations. Divinity is the chief raw material out of which gods have been fashioned. Today we must melt down the gods and refashion the material into new and effective organs of religion, enabling man to exist freely and fully on the spiritual level as well as on the material.

What precise form these new agencies of religious thought will take it is impossible to say in this period of violent transition. But one can make some general prophecies. The central religious hypothesis will certainly be evolution, which by now has been checked against objective fact and has become firmly established as a principle. Evolution is a process, of which we are products, and in which we are active agents. There is no finality about the process, and no automatic or universal progress; but much improvement has occurred in the past, and there could be much further improvement in the future (though there is also the possibility of future failure and regression).

Thus the central long-term concern of religion must be to promote further evolutionary improvement and to realize new possibilities; and this means greater fulfilment by more human individuals and fuller achievement by more human societies.

Human potentialities constitute the world's greatest resource, but at the moment only a tiny fraction of them is being realized. The possibility of tapping and directing these vast resources of human possibility provide the religion of the future with a powerful long-term motive. An equally powerful short-term motive is to ensure the fullest possible development and flowering of individual personalities. In developing a full, deep and rich personality the individual ceases to be a mere cog or cipher, and makes his own particular contribution to evolutionary fulfilment.

In a way most important of all, an evolution-centred religion can no longer be divided off from secular affairs in a separate supernatural compartment, but will interlock with them at every point. The only distinction is that it is concerned with less immediate, less superficial, and therefore more enduring and deeper aspects of existence.

Meanwhile, religious rituals and moral codes will have to be readapted or remodelled. Besides what Nietzsche called the transvaluation of values, we shall need a transfiguration of thought, a new religious terminology and a reformulation of religious ideas and concepts in a new idiom. A humanist religion will have to work out its own rituals and its own basic symbolism.

In place of eternity we shall have to think in terms of enduring process; in place of salvation in terms of attaining the
satisfying states of inner being which combine energy and
peace. There will be no room for petitionary prayer, but
much value in prayer involving aspiration and self-
exploration. A religion of fulfilment must provide bustling
secular man with contacts with all that is permanent and
enduring, with the deeper and higher aspects of existence;
indeed, with every possible opportunity of transcending the
limitations not only of his day-by-day existence in the
equivalents of shared worship, but of his little secular self
in acts of meditation and self-examination and in retreats
from the secular world of affairs. It will of course continue
to celebrate the outstanding events of personal and national
existence (already in some countries there are humanist
wedding and funeral ceremonies). Furthermore, it will enlist
the aid of psychologists and psychiatrists in helping men
and women to explore the depths and heights of their
own inner selves instead of restlessly pursuing external
novelty, to realize more of their mental and spiritual
possibilities, to utilize even their repressed and guilty
urges, and to transcend the limitations and the internal conflicts of the unregenerate self in a constructive wholeness
and a sense of achieving contact or union with a fuller
reality.

Christianity is a universalist and monotheist religion of
salvation. Its consolidation and explosive spread, achieved
through a long period of discussion and zealous ferment,
released vast human forces which have largely shaped the
Western world as we know it. An evolutionary and humanist
religion of fulfilment could be more truly universal and
could release even vaster human forces, which could in large
measure shape the development of the entire world. But its
consolidation and spread will need a period of discussion and
ferment, though with modern communications this is likely
to be much shorter than for Christianity.

The evolutionary vision of man's place and role in the
universe which science and scholarship have given us, could
be the revelation of the new dispensation. Dr Robinson's

article is evidence of its effectiveness in changing ideas. What we now need is a multitude of participants to take part in the great discussion and to join in the search for the larger truth and the more fruitful pattern of belief which we confidently believe is waiting to be elicited.

THE ENLIGHTENMENT
AND THE POPULATION PROBLEM[1]

W HEN I started looking into the matter I found there was
nothing comprehensive on the attitude of the Enlightenment to population, but I have gleaned a good deal
from the *Encyclopédie* itself and still more from the *Philosophical Transactions* of the Royal Society, which we scientists in
England affectionally call the "Phil. Trans.", and from a
rather miscellaneous crop of books[2] with which I will not
weary my readers.

Let me first of all give a little factual background about the
then state of the population and its growth. World population in 1650 was only about 470 million, with a rate of increase of well under one half of 1 per cent a year. A marked
absolute increase began about that time and also an increase
in the rate of growth, doubtless due to the beginning of the
industrial revolution and the great expansion of trade over-
seas and elsewhere; and of course the expansion was particularly rapid in the colonies of the New World. It became
almost explosive—at any rate much more rapid—around
1760 in England, rather later in the rest of Western Europe.
As a result, world population doubled in nearly two centuries,
between 1650 and 1840. It then doubled again in a little
under a century, largely as a result of better public health and
sanitation. Then after that came modern medicine, the result
of which has been called "death-control"—the control of
mortality, especially through the conquest of infectious diseases—and the onset of the real population explosion may be

[1] Part of a paper read to the Conference on the Enlightenment held at
Geneva in July 1963.
[2] For the benefit of continental readers, I would like to cite the following
English and American books which I found useful: Hauser and Duncan's
monumental *The Study of Population*; Wolf's *History of Science and Technology
in the 18th Century*; Bonar's *Theories of Population from Adam Smith to Arthur
Young* (1931); Griffith's *Population Problems of the Age of Malthus*; *World
Population and Resources* (PEP); and David Glass's *Introduction to Malthus*.

dated about 1920. The population of the world will roughly double again by 1980—in a mere sixty years from 1920.

The annual rate of increase only reached 1 per cent during the present century. It is already over 1·75 per cent, and going up; so whatever happens, short of a disastrous atomic war, there will be about 6 billion (American billions, I am glad to say: real billions would mean the end of our human world!) by A.D. 2000. This is well within the lifetime of a good many people now alive. As a result we have had what the Marxian dialectics call the transformation of mere quantity into quality; and most of the resulting qualitative changes have been bad.

To sum up, in the 350 years since the beginning of the Enlightenment there will have been an increase of 5·5 billion as against one of well under 0·5 billion in the 6000 years of previous civilization. So now you see why we speak of a "population explosion".

As regards the Enlightenment, I would like to define it as having been inspired notably by Newton and Locke, and as dating from about 1630 to 1800: that is roughly from the first meetings in London of the so-called Invisible College, which was the forerunner of the Royal Society (whose actual foundation was in 1660), to the date of the first census in Britain, 1801. One point I want to make is the disparity during this period in the absolute numbers and in the rate of increase between the two great rivals of the time, France and Great Britain. In 1650 the population of France was about 19 million and that of Great Britain under 5·5 million. By 1700 the figures were about 20 million and 6·25 million; by 1750, 22 million and 7·5 million; and by 1800, 28 million and rather under 11 million. Thus the disparity, though persistent, was reduced during the period: from 1650 to 1960 the total population of France multiplied itself about 2·5 times, whereas that of Great Britain multiplied itself nearly 10 times.

Historically one must split the consideration of the population problem into several distinct phases. I will begin with what may be called the pre-demographic era, dating from the very end of the sixteenth century to the mid-seventeenth century and the beginning of the Enlightenment proper. Here we find for the first time an interest in human numbers as a subject of more or less rational enquiry, as against purely theological or metaphysical discussion. It began in the Renaissance, partly, of course, as a result of the discovery and colonization of the New World. The overriding background ideas about population in this period were connected with political power, and especially war: first, the need for abundant manpower, especially military manpower (and, of course, high quality healthy manpower), and therefore that an increase of the national population was essentially a good thing; secondly, that over-population was a major cause of wars.

The first man who was really interested in this subject seems to have been a Piedmontese, Botero, who wrote in 1589. He adumbrated Malthus by pointing out that the growth of cities "stops from want of nourishment and support". He thought that world population had reached its limit 3000 years ago—a curious belief.

Then we have Sir Walter Raleigh, one of the greatest polymaths in history. He maintained that war was inevitably caused by the overpopulation of a country. He rightly claimed that countries would not suffer depopulation through wars and through the founding of colonies. As one of the founders of the English colonies in the New World, he was naturally concerned about this point. I was much interested to find that in a curious way he foreshadowed the idea of evolution. He was going into the question of whether Noah's ark would have been big enough to hold all the animals it was said to have held, and he calculated that it would have been big enough, provided that all the species of animals which were contained in it (and by *species* he meant what we would now call *genera* or main kinds) would, after they came out on Mount Ararat, have been able by differentiation or hybridization to produce all varieties now existing—in other words what we now call the separate species. A fascinating adumbration of Darwin!

Of course Bacon was a very important figure in this period. He was interested in population because of its relevance to

229

military and political power. He did not want "great population with little strength": he wanted good yeomen to do our fighting for us. He thought that the enclosures, which were already beginning in England, were leading to a dangerous dearth of people. He was one of the earliest men to concern himself with statistics, but he confined himself to statistics of longevity.

Another important pioneer was Hobbes. Like Raleigh, he thought that overpopulation was a main cause of war and would eventually be the cause of the war to end all wars at the end of the world. He was very much concerned with transplanting the excess of poor yet strong people to new lands. This idea of getting rid of a surplus population was for long a popular one. Today it is no longer practicable, because there are no new lands and vast open spaces to which to export the excess of numbers.

A very interesting man, not so well known as he ought to be, was Sir John Harington, who wrote in 1656. In passing, he seems to have influenced John Adams considerably. He began thinking about changes in the standard of living in relation to population, as well as changes in mere number of people. He was very much concerned with military strength, and thought that Scotland should supply men to England for her wars. He advocated that Ireland, already a trouble spot then, as it has been ever since until recently, should be colonized by Jews. This I do not quite understand. The Jews had, of course, only just been readmitted to England under Oliver Cromwell, so I do not know why Harington wanted to get rid of them again. As was usual in this period, he thought that an increase in population must be a good thing. He therefore proposed that fathers of ten children should pay no taxes, with double taxes for bachelors over 25.

When we get to the Enlightenment proper we find that governments are beginning to take an interest in population, or at least in certain classes of the population: householders, taxpayers and, above all, men of military age useful for war—in the language of the time, good "fencible men". They instituted surveys to ascertain the number of these various classes of the population, but they were very sporadic and not well planned; they were also often evaded because, naturally, people did not want to be taxed or taken for military service. There was no census, largely because of the theological prejudice and religious feeling, according to Biblical precedent, against "numbering the people". I would remind you that this prejudice had serious practical consequences. A Bill for taking annual censuses in Britain was actually passed by the House of Commons in 1754 but was thrown out by the Lords, because, they said, it would lead to unrest among the lower classes, who would regard it as contrary to divine will. So we had no proper census in Britain until 1801. Meanwhile, however, there was a real scientific interest in population as a subject of "useful knowledge", and there are a number of papers on it in the Royal Society's *Philosophical Transactions*. This led to new methods (which I shall deal with later) for estimating the true size of the population.

With this we pass from the pre- to the proto-demographic era. In this phase the pioneers were, again, two Englishmen: John Graunt and Sir William Petty.

Graunt discovered the value of the existing statistical records, in the shape of the so-called bills of mortality, and his results were published in 1661 in the Royal Society's *Transactions*. It was the plague which really stimulated Graunt's work. Indeed, the idea of making up for the loss of population due to the plague had actually led to proposals being made, not by Graunt but by other people, for introducing polygamy in Britain![1] Graunt really is the pioneer of "speaking in numbers". He thinks numerically; he has a really quantitative approach. He was the first to think of constructing life-tables, and he correlated incidence of disease with death-rate. As a result of his work the city of Paris, a few years later, started using bills of mortality to estimate its population.

Petty was a true inventor. He too based his study on bills

[1] Those interested in this curious subject will find an excellent account in Dr Aldridge's article on "Population and polygamy in eighteenth century thought", *Journal of the History of Medicine* (1949), iv. 129.

of mortality, in Dublin and elsewhere. He believed, again, that increased size of population, whether of the nation or of its capital city, was a good thing, and he proclaimed that "fewness of people is the real poverty". He asserted that a nation of 8 million people was more than twice as rich as one with 4 million, thus linking population with economics. He, again, advocated the transplantation of people. For instance he wanted to transplant three-quarters of the inhabitants of the Scottish Highlands to the Lowlands, where he thought they would be more useful. In a later work he proved, to his great satisfaction as an Englishman, that London was the biggest city in the world. He makes numerous calculations: "It is equivalent to Paris and Rouen together. It is equivalent to Amsterdam, Venice and Rome together. It is equivalent to any four French cities together."

All these studies, aided by the rapid development of mathematics, led to really good life-tables, largely for insurance purposes. They also led to early theories of probability, thus foreshadowing one of the great methodological inventions of our scientific age.

Dr Besterman has confirmed my supposition that this turned out to be of considerable practical importance to Voltaire. One day, at a gathering of *philosophes* and their friends, the mathematician La Condamine pointed out that the basis of an official lottery which had just been announced to liquidate certain annuities was unsound, and that by buying up all the tickets one would make 3 million francs profit. Everybody was interested, but Voltaire was the only one present who took practical steps. He rushed off and organized a stock company for the purpose, and himself made a quarter of a million francs profit as a result.

We must remember, of course, that all this took place against a social, political and ideological background extraordinarily different from that of the present century. This is well illustrated by Mirabeau's treatise of 1756, *L'ami des hommes ou traité sur la population*. As I have already mentioned, one element in the eighteenth-century background was a major preoccupation with military power and consequently with increasing the population, notably in little

England as against great big France. In the second place, the subject was linked with the problem of colonies, and with the mercantilist view that colonies were to be established primarily for the benefit of the mother country. Accordingly there was great discussion as to whether emigration to the colonies would adversely affect a country's population strength as well as its economic strength. There was also great discussion as to whether industrialization, especially urbanization, would lead to depopulation, especially depopulation of the countryside as against the city.

Then there was all the time the question: what was keeping population-increase down? The chief blame fell on "vice". The outstanding vice in eighteenth-century England, of course, was gin. A second was promiscuity and consequent venereal diseases. There was also a great deal of discussion as to what sort of government would promote population-increase; and this, of course, was discussed against those rather abstract and perfectionist ideas about human nature, inevitable progress, social possibilities, and the power of reason and education, which were the keynotes of the Enlightenment. For the *philosophes*, there would inevitably be a rush of progress once the *infâme* of bigotry and tyranny had been crushed. This over-optimistic, almost millenary outlook culminated in works like Godwin's *Political Justice*, in 1793, and Condorcet's *Esquisse d'un tableau historique du progrès de l'esprit humain* a few years later.

On the side-lines, there was a curious controversy as to whether the ancient or the modern world had a larger population, and this bothered people's minds to a quite extraordinary degree. A notable contributor to this discussion was Hume, with his essay on *The Populousness of Ancient Nations*. Unlike Gibbon, who estimated that the Roman Empire in A.D. 50 had a population of 120 million, Hume thought quite rightly that the ancient world was smaller than the modern. Montesquieu outdid Gibbon by asserting that the population of the eighteenth century world was only about one-tenth of that of the ancient world, and was very properly rebuked by Voltaire. In a way, this discussion was the demographic equivalent of the Battle of the Books,

233

bedevilled by belief that the scriptures were literally true and therefore that an extremely short time had elapsed since the Creation, and further that there would be a similarly short span of time before the last judgment and the end of the world.

The replacement of the eighteenth century's abstract, idealistic, and one might almost say scholastic approach to population by a scientific one was first adumbrated, so far as I can see, by Quesnay in the 1750's, by Adam Smith in the *Wealth of Nations* in 1776, and effectively brought about once and for all by Malthus around 1800.

Now if I may I will bring up a few particular examples. The first thing I did was to look at the article "Population" in the *Encyclopédie*, first edition. It is by Damilaville, a friend of Voltaire. He begins with a reference to the elaborate calculations of Robert Wallace, an Englishman, about the possible increase from the single couple, Adam and Eve, in the 1233 years between the Creation and the Deluge. He estimated that this would give rather over 412 "real" billions—million millions. Of course Voltaire made fun of this sort of thing. He calculated what would have been produced by the three sons of Noah and their wives in the next few centuries. This was deliberately intended to ridicule people like Wallace and the Jesuit Peteau, as "ceux qui font des enfants à coup de plume" by their unrealistic calculations.

Damilaville concluded that the causes of increase or of decrease of population were infinite, as against Malthus's more scientific conclusion that there was only one essential cause, namely, the relation of population to resources. In passing, it is worth recalling that something very similar happened in biology when Darwin replaced the many conflicting ideas about what caused adaptation and evolution by the single causal factor of natural selection.

Damilaville also came to the conclusion that the total number of men inhabiting the earth "has been, is and will always be about the same for all time". In spite of this, he goes on to discuss what sort of social institutions would promote increase.

Buffon, that very distinguished but politically and ideo-
logically cautious scientist, touched on the problem in 1749,
in the second volume of his great *Histoire Naturelle*. In the
chapter entitled "Histoire naturelle de l'homme", after pointing out that man's life-span is much more variable than that
of other mammals, and citing the work of Graunt, Halley,
Kersboom and others, he attempts to obtain figures of human
mortality at different ages, with a view to ascertaining the
"degrees of probability of length of life"—in other words, a
life-expectancy table. On the basis of Dupré de Saint-Maur's
figures from three Parisian and twelve country parishes, he
prints such a table, which gives 7 as the age at which the
expectancy of further life is greatest (42¼ years). His final
comments are typical of the general attitude of the eighteenth-
century *philosophes:*

> Ces vérités physiques si mortifiantes en elles-mêmes peuvent se
> compenser par des considérations morales; un homme doit regarder
> comme nulles les 15 premières années de sa vie. Tout ce qui s'est
> passé dans ce long intervalle de temps est effacé de sa mémoire . . .
> ce n'est pas la même succession d'idées, ni, pour ainsi dire, la même
> vie; nous ne commençons à vivre moralement que quand nous
> commençons à ordonner nos pensées, à les tourner vers un certain
> avenir, et à prendre une espèce de consistance, un état relatif à ce
> que nous devons être dans la suite.
>
> En considérant la durée de la vie sous ce point de vue, *qui est le
> plus réel* (my italics), nous trouverons dans la table qu'à l'âge de
> 25 ans on n'a vécu que le quart de la vie, qu'à l'âge de 38 ans
> on n'en a vécu que la moitié, et que ce n'est qu'à l'age de 56 ans
> qu'on a vécu les trois quarts de sa vie.

There is a rather interesting contribution from an English
parson, the Reverend William Brakenridge, in 1750. He
calculated that "the earth could support 20 times the present
number; therefore it will be over a thousand years before it
can be fully peopled". He concludes that "the origin of mankind is not more ancient than commonly supposed"—which,
of course, was 4004 B.C. On the practical side he came to the
quite false conclusion that the population of England was

decreasing and could only be maintained by "supplies" (to use his word) from Scotland and Ireland.

He was countered by another parson, the reverend Mr Foster, also in the "Phil. Trans". He concluded that the population of England was increasing, and he wanted to use every possible means of stimulating this increase, including the naturalization of aliens.

At about the same time—to be precise, in 1757—Benjamin Franklin had written for the *London Observer* concerning the increase of mankind, drawing attention to the ample space and new resources of the American colonies as against their limitations in England. He influenced Malthus, and he also influenced the Reverend Richard Price (another clergyman) who wrote a so-called "Letter to Benjamin Franklin", also published in the "Phil. Trans.", to which he gave the charming miscellaneous title "On the expansion of lives, the increase of mankind, and the population of London". He gives an early definition of life expectancy—a very interesting new venture. He was the first man to bring out the idea of the rate of doubling of population, and relate it to resources. From his available statistics he concluded that the population of Madeira doubled in 84 years, that of the American colonies in general doubled in 25 years, and that of the back settlements of America in 15 years.

He prophesied that the white population of North America would by 1830 be twice that of Great Britain. (I do not know what the actual figures are but I should think that the increase was a good deal more.) He found that life expectancy was much lower in London—only about 20 years, as against about 39 in Madeira—and he asserted (quite rightly for the period) that all cities had to be replenished by immigration from the countryside. He said this applied even to cities like Boston in the New World. "London alone," he said, "is a gulph which swallows up an annual increase equal to nearly three-quarters of that of Sweden." This implies the relative depopulation of the countryside. But Oliver Goldsmith, inspired by Price, took it into his head that it meant an absolute depopulation, and was prompted to write, in 1770, *The Deserted Village*, one of the few good poems mainly concerned with the subject of population. I think it is worth while quoting some of his remarkable lines. It was in the first place a protest against luxury:

> Ten thousand arts combined
> To pamper luxury and thin mankind.
> . . . .
> Ill fares the land, to hastening ills a prey,
> Whose wealth accumulates and men decay.

He also protested against enclosures, and against the owners of the new big country houses and parks who were buying up all the land:

> One only master grasps the whole domain.
> . . . The man of wealth and pride
> Takes up the space that many poor supplied.

Until eventually:

> . . . Now the sounds of population fail.
> No cheerful murmurs fluctuate in the gale.

The *Deserted Village* is a very interesting historical document.

Almost all the French *philosophes* deal with the subject of population, as far as I can make out, in relation to the assumed needs of increasing it. They deal with the problem as a means of airing their general social and political views, on the need to abolish intolerance and arbitrary political powers, the possibility of human perfection and unlimited progress, and so on.

I will end with a little comic relief from the Maréchal de Saxe. He may have been a military genius, but he was not much of a *philosophe*. He supported the view that there had been a marked decrease in human numbers since the time of Julius Caesar, and he quoted another military genius, Maréchal Vauban, as assessing the population of France only 50 years earlier at 20 million—which was, of course, quite ridiculous. He believed that we must try to reverse this trend by every means. If we do not obey the first of God's commandments, to increase and multiply, "la faculté d'engendrer se perdra"—an anticipation of the biological errors of another

237

Frenchman of genius, Lamarck. Human propagation, he sweepingly concludes, is chiefly opposed by education, and by marriage, which is "against the laws of nature"(!) He proposed an elaborate system of temporary but renewable marriages—unions which should be for propagation only. A woman should have the right to choose her sexual partner, and he laid down the general statement that "the more children a woman has the happier she is". I do not think that women today would say that this was universally true. It reminds me of Stevenson's famous lines,

> The world is so full of a number of things
> I am sure we should all be as happy as kings;

and Thurber's comment—"we all know how happy kings are". He ends: "If one woman produced an average of six children, in 108 years a million women would produce 978 million human beings. This number is enormous. . . ."

The period closes with Malthus. His *Essay on Population* (which, by the way, was published anonymously because he thought it would be a cause of scandal for a clergyman to deal with these matters) was first published in 1798. The second edition, five years later, very much revised, bore his name. Much later, in 1830, he summed up his whole doctrine with his *Summary View of the Principle of Population*. Note this word "principle". It is very important, for here with Malthus we pass from considering only demographic techniques combined with academic speculations on the population of the ancient world, and with rather Utopian speculations on how to increase population, to a scientific approach to a real population theory and its principles—very much as with Darwin in the field of evolution. And do not let us forget that both Darwin and Wallace got their idea of natural selection from him. One of my favourite quotations is from Darwin's autobiography. He had realized that artificial selection could be effective in transforming man's domestic animals and crop plants, but could not see what agency might be at work in the same way in nature, until one day, "happening to read the Reverend Mr Malthus on *Population* for amusement, the solution flashed into my head". I find it charming to read Malthus for amusement.

Malthus deduced everything from certain well-defined principles. First, the tendency for population to increase geometrically, by compound interest. (Note the word *tendency*. This was one of the first occasions on which scientific consideration was given to inherent tendency. This is extremely important. For instance, in relation to the disputes about heredity and environment, instead of genetic determination, we now think in terms of genetic tendencies.) But, Malthus continued, whereas population tends to increase geometrically, food, the ultimate resource, has no such inherent tendency, and can never do this. (Malthus made a mistake in saying food could only increase arithmetically, but that is a minor error.) As a result, there is a tendency for population to outrun subsistence.

Then, as a further result, there must be checks operating to keep the increase down. These checks he summarized in a curious phrase: "Vice, misery and moral restraint." By "vice", he meant gin, promiscuity and venereal disease; by "misery", starvation and disease; by "restraint", the avoidance of early marriage and abstinence from sexual intercourse. Again like Darwin, he amassed an enormous number of facts relevant to the problem—about the relation of social conditions to population-increase, facts about rates of doubling, about the slowing down of food-production—and drew conclusions from them.

These principles have remained basic to all studies of population since then, as have Darwin's principles in all studies of evolution since his time. On the other hand, the conclusions Malthus drew from them are very different from ours today. For instance, he had very strong views about the dangers of poor relief. He believed that both private property and poverty were inevitable. He was hostile to deliberate birth-control. He had no real idea of what democracy meant, or could mean, and was much influenced by the prevalent British fear of revolution. But he was one of the first to be interested in the economic consequences and implications of population-increase. Although he proclaimed himself a rationalist, a believer in reason and intelligent restraint, all

239

the same his attitude was basically non-rational and empirical.

Malthus thus marks the beginning of the modern era in demography. From now on men look at population-growth as a process in time, with rate of increase as its central feature; with the inevitable conclusion that we must eventually work towards the control of rate of increase, so as to secure an optimum balance between population and resources. These ideas have very important implications at the present moment, for instance in regard to the pressing problem of industrializing densely populated and underdeveloped countries; because if too many babies are born, their feeding and educating and servicing takes up so much capital and skill that not enough is left to transform the economy. But the whole question of population has now assumed a new and alarming significance for the human race, and in the essay that follows I shall consider some of the problems with which it faces us.

THE CROWDED WORLD

POPULATION has at last made the grade and emerged as a World Problem. Unfortunately, most of those who speak or write about the problem persist in thinking of it in terms of a race between human numbers and world resources, especially of food—a kind of competition between production and reproduction. The neo-Malthusians, supported by progressive opinion in the Western World and by leading figures in most Asian countries, produce volumes of alarming statistics about the world population explosion and the urgent need for birth-control, while the anti-Malthusians, supported by the two ideological blocs of Catholicism and Communism, produce equal volumes of hopeful statistics, or perhaps one should say of wishful estimates, purporting to show how the problem can be solved by science, by the exploitation of the Amazon or the Arctic, by better distribution, or even by shipping our surplus population to other planets.

Certainly, the statistics are important. The major fact emerging from them is that there really *is* a population explosion. During the millennia of man's early history, world population increased steadily but very slowly, so that by the end of the seventeenth century it had barely topped the half-billion mark. But then, as a result of the great explorations during and after the Renaissance, and still more of the rise of natural science and technology at the end of the seventeenth century, the process was stepped up, so that by the beginning of the present century world population stood at about $1\frac{1}{4}$ billion, and its compound interest rate of increase had itself increased from under $\frac{1}{2}$ of 1 per cent in 1650 to nearly 1 per cent (and we all know what big results can flow from even a small increase in compound interest rates).

But the real explosion is a twentieth-century phenomenon, due primarily to the spectacular developments in medicine and hygiene, which have drastically cut down death-rates

241

without any corresponding reduction in birth-rates—death-control without birth-control. The compound interest rate of increase meanwhile crept, or rather leapt, up and up, from under 1 per cent in 1900 to 1½ per cent at mid-century, and nearly 1¾ per cent today; and it will certainly go on increasing for some decades more. This means that the *rate* of human doubling has itself been doubled within the past 80 years. World population has more than doubled since 1900 to reach about 2¾ billion today; and it will certainly reach well over 5½ billion, probably 6 billion, and possibly nearly 7 billion by the year 2000.

Coming down to details, Britons will be jolted by the fact that the net increase of world population amounts to about 150,000 every 24 hours, or the equivalent of a good-sized New Town every day—Hemel Hempstead yesterday, Harlow today, Crawley tomorrow, and so on through the weeks and months; while Americans will be startled out of any complacency they may have possessed by the fact that this is the equivalent of 10 baseball teams, complete with coach, every minute of every day and night. Such facts make the idea of interplanetary disposal of the earth's surplus population merely ridiculous.

It is also salutary to be reminded that the number of human beings alive in A.D. 1999—within the lifetime of many now living—will be about double that of those alive today; that some populations, like that of Barbados, are growing at a rate of over 3 per cent. compound interest per annum, which means doubling in less than 20 years; that in an underdeveloped but already densely populated country like India, successful industrialization will be impossible unless the birth-rate is cut to about half within the next 30 or 40 years, for otherwise the capital and the trained man- and woman-power needed to give the country a stable industrial economy will be swallowed up in feeding, housing, educating, and servicing the excess population; that religious opposition to population-control is strongest and most effective in regions like Latin America, where population-increase is most rampant; that there is no provision for international study and research on population-control as there is on atomic energy, on the world's arid zones, on brain function, or on oceanography; that there is already an alarming (and increasing) shortage of available water-supplies, high-grade mineral resources, and educational facilities, even in industrially advanced countries like the U.S.A.; that the annual increase of Communist China's population is 13 million, more than the equivalent of a new Sweden and a new Denmark every year; or that the World Health Organization has twice been prevented by Roman Catholic pressure from even considering population-density as a factor in the world's health.

But in the broad view the most important thing about the population explosion is that it is making everyone—or rather everyone capable of serious thought—think seriously about the future of our human species on our human planet.

The Middle Ages were brought to an end by a major revolution in thought and belief, which stimulated the growth of science and the secularization of life at the expense of significance in art and religion, generated the industrial-technological revolution, with its stress on economics and quantitative production at the expense of significance in quality, human values and fulfilment, and culminated in what we are pleased to call the Atomic Age, with two World Wars behind it, the threat of annihilation before it, and an ideological split at its core.

Actually our modern age merits the adjective atomistic rather than atomic. Further, it will soon become very unmodern. For we are on the threshold of another major revolution, involving a new pattern of thought and a new approach to human destiny and its practical problems. It will usher in a new phase of human history, which I like to call the Evolutionary Age, because it envisages man as both product and agent of the evolutionary process on this planet.

The new approach is beginning to take effect in two rather distinct fields, of ecology and ideology, and is generating two parallel but linked currents of thought and action, that may be called the Ecological Revolution and the Humanist Revolution.

The population explosion is giving a powerful impetus to

243

both these revolutionary currents. Ecology is the science of relational adjustment—the balanced relations of living organisms with their environment and with each other. It started botanically in a rather modest way as a study of plant communities in different habitats; went on to the fruitful idea of the ecological succession of different plant communities in a given habitat, leading up to an optimum climax community—mixed forest in the humid tropics, rich grassland on the prairies; was extended to take in animal communities, and so to the illuminating concepts of food-chains and adaptive niches; and finally, though rather grudgingly, was still further enlarged to include human as well as biological ecology.

The population explosion has brought us up against a number of tough ecological facts. Man is at last pressing hard on his spatial environment—there is little leeway left for his colonization of new areas of the world's surface. He is pressing hard on his resources, notably non-renewable but also renewable resources. As Professor Harrison Brown has so frighteningly made clear in his book, *The Challenge of Man's Future*, ever-increasing consumption by an ever-increasing number of human beings will lead in a very few generations to the exhaustion of all easily exploitable fossil fuels and high-grade mineral ores, to the taking up of all first-rate agricultural land, and so to the invasion of more and more second-rate marginal land for agriculture. In fact, we are well on our way to ruining our material habitat. But we are beginning to ruin our own spiritual and mental habitat also. Not content with destroying or squandering our resources of material things, we are beginning to destroy the resources of true enjoyment—spiritual, aesthetic, intellectual, emotional. We are spreading great masses of human habitation over the face of the land, neither cities nor suburbs nor towns nor villages, just a vast mass of urban sprawl or subtopia. And to escape from this, people are spilling out farther and farther into the wilder parts and so destroying them. And we are making our cities so big as to be monstrous, so big that they are becoming impossible to live in. Just as there is a maximum possible size for an efficient land animal—you can't have a land

animal more than about twice as large as an elephant—
so there is a maximum possible efficient size for a city.
London, New York, and Tokyo have already got beyond
that size.

In spite of all that science and technology can do, world
food-production is not keeping up with world population,
and the gap between the haves and the have-nots of this
world is widening instead of being narrowed.

Meanwhile everywhere, though especially in the so-called
Free Enterprise areas of the world, economic practice (and
sometimes economic theory) is concerned not primarily with
increased production, still less with a truly balanced economy, but with exploitation of resources in the interests of
maximized and indiscriminate consumption, even if this
involves present waste and future shortage.

Clearly this self-defeating, self-destroying process must be
stopped. The population explosion has helped to take our
economic blinkers off and has shown us the gross and increasing imbalance between the world's human population
and its material resources. Unless we quickly set about
achieving some sort of balance between reproduction and
production, we shall be dooming our grandchildren and all
their descendants, through thousands upon thousands of
monotonous generations, to an extremely unpleasant and
unsatisfactory existence, overworked and undernourished,
overcrowded and unfulfilled.

To stop the process means planned conservation in place
of reckless exploitation, regulation and control of human
numbers, as well as of industrial and technological enterprise, in place of uninhibited expansion. And this means an
ecological approach. Ecology will become the basic science
of the new age, with physics and chemistry and technology
as its hand-maidens, not its masters. The aim will be to
achieve a balanced relation between man and nature, an
equilibrium between human needs and world resources.

The Humanist Revolution, on the other hand, is destined
to supersede the current pattern of ideas and beliefs about
nature (including human nature) and man's place and role
in it, with a new vision of reality more in harmony with man's

present knowledge and circumstances. This new pattern of ideas can be called humanist, since it is focused on man as a product of natural evolution, not on the vast inanimate cosmos, nor on a God or gods, nor on some unchanging spiritual Absolute. For humanism in this sense, man's duty and destiny is to be the spearhead and creative agent of the over-all evolutionary process on this planet.

The explosive growth of scientific and historical knowledge in the past hundred years, especially about biological and human evolution, coupled with the rise of rationalist criticism of established theologies and ancient philosophies, had cleared the ground for this revolution in thought and executed some of the necessary demolition work. But now the population explosion poses the world with the fundamental question of human destiny—*What are people for?* Surely people do not exist just to provide bomb-fodder for an atomic bonfire, or religion-fodder for rival churches, or cannon-fodder for rival nations, or disease-fodder for rival parasites, or labour-fodder for rival economic systems, or ideology-fodder for rival political systems, or even consumer-fodder for profit-making systems. It cannot be their aim just to eat, drink and be merry, and to hell with posterity. Nor merely to prepare for some rather shadowy after-life. It cannot be their destiny to exist in ever larger megalopolitan sprawls, cut off from contact with nature and from the sense of human community and condemned to increasing frustration, noise, mechanical routine, traffic congestion and endless commuting; nor to live out their undernourished lives in some squalid Asian or African village.

When we try to think in more general terms it is clear that the dominant aim of human destiny cannot be anything so banal as just maximum quantity, whether of human beings, machines, works of art, consumer goods, political power, or anything else. Man's dominant aim must be increase in quality—quality of human personality, of achievement, of works of art and craftsmanship, of inner experience, of quality of life and living in general.

"Fulfilment" is probably the embracing word: more fulfilment and less frustration for more human beings. We want more varied and fuller achievement in human societies, as against drabness and shrinkage. We want more variety as against monotony. We want more enjoyment and less suffering. We want more beauty and less ugliness. We want more adventure and disciplined freedom, as against routine and slavishness. We want more knowledge, more interest, more wonder, as against ignorance and apathy.

We want more sense of participation in something enduring and worth while, some embracing project, as against a competitive rat-race, whether with the Russians or our neighbours on the next street. In the most general terms, we want more transcendence of self in the fruitful development of personality: and we want more human dignity not only as against human degradation, but as against more self-imprisonment in the human ego or more escapism. But the inordinate growth of human numbers bars the way to any such desirable revolution, and produces increasing frustration instead of greater fulfilment.

There are many urgent special problems which the population explosion is raising—how to provide the increasing numbers of human beings with their basic quotas of food and shelter, raw materials and energy, health and education, with opportunities for adventure and meditation, for contact with nature and with art, for useful work and fruitful leisure; how to prevent frustration exploding into violence or subsiding into apathy; how to avoid unplanned chaos on the one hand and over-organized authoritarianism on the other.

Behind them all, the long-term general problem remains. Before the human species can settle down to any constructive planning of his future on earth (which, let us remember, is likely to be many times longer than his past, to be reckoned in hundreds of millions of years instead of the hundreds of thousands of his prehistory or the mere millennia of History), it must clear the world's decks for action. If man is not to become the planet's cancer instead of its partner and guide, the threatening plethora of the unborn must be for ever banished from the scene.

Above all we need a world population policy—not at some

unspecified date in the future, but now. The time has now come to think seriously about population policy. We want every country to have a population policy, just as it has an economic policy or a foreign policy. We want the United Nations to have a population policy. We want all the international agencies of the U.N. to have a population policy.

When I say a population policy, I don't mean that anybody is going to tell every woman how many children she may have, any more than a country which has an economic policy will say how much money an individual businessman is allowed to make and exactly how he should do it. It means that you recognize population as a major problem of national life, that you have a general aim in regard to it, and that you try to devise methods for realizing this aim. And if you have an international population policy, again it doesn't mean dictating to backward countries or anything of that sort; it means not depriving them of the right (which I should assert is a fundamental human right) to scientific information on birth-control, and it means help in regulating and controlling their increase and planning their families.

Its first aim must be to cut down the present excessive rate of increase to manageable proportions: once this is done we can think about planning for an optimum size of world population—which will almost certainly prove to be less than its present total. Meanwhile we, the people of all nations, through the U.N. and its Agencies, through our own national policies and institutions, and through private Foundations, can help those courageous countries which have already launched a population policy of their own, or want to do so, by freely giving advice and assistance and by promoting research on the largest scale.

When it comes to United Nations agencies, one of the great scandals of the present century is that owing to pressure, mainly from Roman Catholic countries, the World Health Organization has not been allowed even to consider the effects of population density on health. It is essential and urgent that this should be reversed.

248

There is great frustration in the minds of medical men all over the world, especially those interested in international affairs, who, at the cost of much devoted labour, have succeeded in giving people information on how to control or avoid disease. Malaria in Ceylon is a striking example. As a result of all this wonderful scientific effort and goodwill, population has exploded, and new diseases, new frustrations, new miseries are arising. Meanwhile medical men are not allowed to try to cope with these new troubles on an international scale—and indeed sometimes not even on a national scale. It is an astonishing and depressing fact that even in the advanced and civilized U.S.A. there are two States in which the giving of birth-control information on medical grounds even by non-Catholic doctors to non-Catholic patients, is illegal.

In conclusion I would simply like to go back to where I started and repeat that we must look at the whole question of population increase not merely as an immediate problem to be dealt with *ad hoc*. We must look at it in the light of the new vision of human destiny which human science and learning has revealed to us. We must look at it in the light of the glorious possibilities that are still latent in man, not merely in the light of the obvious fact that the world could be made a little better than it is. We must also look at it in the light of the appalling possibilities for evil and misery that still hang over the future of evolving man.

This vision of the possibilities of wonder and more fruitful fulfilment on the one hand as against frustration and increasing misery and regimentation on the other is the twentieth-century equivalent of the traditional Christian view of salvation as against damnation. I would indeed say that this new point of view that we are reaching, the vision of evolutionary humanism, is essentially a religious one, and that we can and should devote ourselves with truly religious devotion to the cause of ensuring greater fulfilment for the human race in its future destiny. And this involves a furious and concerted attack on the problem of population; for the control of population is, I am quite certain, a prerequisite for any radical improvement in the human lot.

We do indeed need a World Population Policy. We have learnt how to control the forces of outer nature. If we fail to control the forces of our own reproduction, the human race will be sunk in a flood of struggling people, and we, its present representatives, will be conniving at its future disaster.

EUGENICS IN EVOLUTIONARY
PERSPECTIVE[1]

How does eugenics look in our new evolutionary perspective? Man, like all other existing organisms, is as old as
life. His evolution has taken close on three billion years.
During that immense period he—the line of living substance leading to *Homo sapiens*—has passed through a series
of increasingly high levels of organization. His organization
has been progressively improved, to use Darwin's phrase,
from some submicroscopic gene-like state, through a uni-
cellular to a two-layered and a metazoan stage, to a three-
layered type with many organ-systems, including a central
nervous system and simple brain, on to a chordate with
notochord and gill-slits, to a jawless and limbless vertebrate,
to a fish, then to an amphibian, a reptile, an unspecialized
insectivorous mammal, a lemuroid, a monkey with much
improved vision, heightened exploratory urge and manipul-
ative ability, an ape-like creature, and finally to a fully human
organism, big-brained and capable of true speech.

This astonishing process of continuous advance and biological improvement has been brought about through the
operation of natural selection—the differential reproduction
of biologically beneficial combinations of mutant genes,
leading to the persistence, improvement and multiplication
of some strains, species and patterns of organization and the
reduction and extinction of others, notably to a succession of
so-called dominant types, each achieving a highly successful
new level of organization and causing the reduction of previous dominant types inhabiting the same environment.

In biologically recent times, one primate line broke
through from the mammalian to the human type of organization. With this, the evolutionary process passed a critical
point, and entered on a new state or phase, the psychosocial,
differing radically from the biological in its mechanism, its

[1] The Galton Lecture, delivered in London on 6 June 1962.

tempo, and its results. As a result, man has become the latest dominant type in the evolutionary process, has multiplied enormously, has achieved miracles of cultural evolution, has reduced or extinguished many other species, and has radically affected the ecology and indeed the whole evolutionary process of our planet. Yet he is a highly imperfect creature. He carries a heavy burden of genetic defects and imperfections. As a psychosocial organism, he has not undergone much improvement. Indeed man is still very much an unfinished type, who clearly has actualized only a small fraction of his human potentialities. In addition, his genetic deterioration is being rendered probable by his social set-up, and definitely being promoted by atomic fallout. Furthermore, his economic, technical and cultural progress is threatened by the high rate of increase of world population.

The obverse of man's actual and potential further defectiveness is the vast extent of his possible future improvement. To effect this, he must first of all check the processes making for genetic deterioration. This means reducing man-made radiation to a minimum, discouraging genetically defective or inferior types from breeding, reducing human over-multiplication in general and the high differential fertility of various regions, nations and classes in particular. Then he can proceed to the much more important task of positive improvement. In the not too distant future the fuller realization of possibilities will inevitably come to provide the main motive for man's over-all efforts; and a Science of Evolutionary Possibilities, which today is merely adumbrated, will provide a firm basis for these efforts. Eugenics can make an important contribution to man's further evolution: but it can only do so if it considers itself as one branch of that new nascent science, and fearlessly explores all the possibilities that are open to it.

*   *   *

In the last twenty-five years many events have occurred and many discoveries have been made with a bearing on eugenics. Events such as the explosion of atomic and nuclear bombs, the equally sinister "population explosion", the *reductio ad horrendum* of racism by Nazi Germany, and the introduction of artificial insemination for animals and human beings, sometimes with the use of deep-frozen sperm; scientific discoveries such as that of DNA (desoxyribonucleic acid) as the self-copying chemical structure which provides the essential basis for heredity and evolution, of subgenic organization, of the widespread existence of balanced polymorphic genetic systems, and of the intensity and efficacy of selection in nature; the realization that the entities which evolve are populations of phenotypes, with consequent emphasis on population-genetics on the one hand, and on the interaction between genotype and environment on the other; and finally the recognition that adaptation and biological improvement are universal phenomena in life. I do not propose to discuss these changes and discoveries now, but shall plunge directly into my subject—Eugenics in Evolutionary Perspective.

The evolutionary perspective includes the broad background of the cosmic past. It is against this background that we must face the problems of the present and the challenge of the future. Let me begin by reiterating that man is an exceedingly recent phenomenon. The earliest creatures which can properly be called men, though they must be assigned to different genera from ourselves, date from less than two million years ago, and our own species, *Homo sapiens*, from much less than half a million years. Man began to put his toe tentatively across the critical threshold leading towards evolutionary dominance perhaps a quarter of a million years ago, but took tens of thousands of years to cross it, emerging as a dominant type only during the last main glaciation, probably later than 100,000 B.C., but not becoming fully dominant until the discovery of agriculture and stock-breeding well under 10,000 years ago, and overwhelmingly so with the invention of machines and writing and organized civilization a bare five millennia before the present day, when his dominance has become so hubristic as to threaten his own future.

All new dominant types begin their career in a crude and imperfect form, which then needs drastic polishing and improvement before it can reveal its full potentialities and achieve full evolutionary success. Man is no exception to this rule. He is not merely exceedingly young; he is also exceedingly imperfect, an unfinished and often botched product of evolutionary improvisation. Teilhard de Chardin has called the transformation of an anthropoid into a man *hominization*: it might be more accurately, though more clumsily, termed *psychosocialization*. But whatever we call it, the process of hominization is very far from complete; the serious study of its course, its mechanisms and its techniques has scarcely started; and only a fraction of its potential results have been realized. Man, in fact, is in urgent need of further improvement.

This is where eugenics comes into the picture. For though the psychosocial system in and by which man carries on his existence could obviously be enormously improved with great benefit to its members, the same is also true for his genetic outfit.

Severe and primarily genetic disabilities like haemophilia, colour-blindness, mongolism, some kinds of sexual deviation, some mental defect, sickle-cell anaemia, some forms of dwarfism, and Huntington's chorea are the source of much individual distress, and their reduction would remove a considerable burden from suffering humanity. But these are occasional abnormalities. Quantitatively their total effect is insignificant in comparison with the massive imperfection of man as a species, and their reduction appears as a minor operation in comparison with the large-scale positive possibilities of all-round general improvement.

Take first the problem of intelligence. It is to man's higher level of intelligence that he owes his evolutionary dominance; and yet how low that level still remains! It is now well established that the human I.Q., when properly assayed, is largely a measure of genetic endowment. Consider the difference in brain-power between the hordes of average men and women with I.Q.s around 100 and the meagre company of Terman's so-called geniuses with I.Q.s of 160 or over, and the much rarer true geniuses like Newton and Darwin, Tolstoy and Shakespeare, Goya and Michelangelo, Hammurabi and Confucius; and then reflect that, since the frequency curve for intelligence is approximately symmetrical, there are as many stupider people with I.Q.s below 100 as there are abler ones with I.Q.s above it.

Recollect also that the great and striking advances in human affairs, as much in creative art and political and military leadership as in scientific discovery and invention, are primarily due to a few exceptionally gifted individuals. Remember that on the established principles of genetics a small raising of average capacity would of necessity result in an upward shifting of the entire frequency curve, and therefore a considerable increase in the absolute numbers of such highly intelligent and well-endowed human beings that form the uppermost section of the curve (as well as a decrease in the numbers of highly stupid and feebly endowed individuals at the lower end).

Reflect further on the fact, originally pointed out by Galton, that there is already a shortage of brains capable of dealing with the complexities of modern administration, technology and planning, and that with the inevitable increase of our social and technical complexity, the greater will that shortage become. It is thus clear that for any major advance in national and international efficiency we cannot depend on haphazard tinkering with social or political symptoms or *ad hoc* patching up of the world's political machinery, or even on improving general education, but must rely increasingly on raising the genetic level of man's intellectual and practical abilities. As I shall later point out, artificial insemination by selected donors could bring about such a result in practice.

The same applies everywhere in the psychosocial process. For more and better scientists, we need the raising of the genetic level of exploratory curiosity or whatever it is that underlies single-minded investigation of the unknown and the discovery of novel facts and ideas; for more and better artists and writers, we need the raising of the genetic level of disciplined creative imagination; for more and better statesmen, that of the capacity to see social and political situations as wholes, to take long-term and total instead of only short-term and partial views; for more and better technologists and engineers, that of the passion and capacity to understand how things work and to make them work more efficiently; for more and better saints and moral leaders, that of disciplined valuation, of devotion and duty, and of the capacity to love; and for more and better leaders of thought and guides of action we need a raising of the capacity of man's vision and imagination, to form a comprehensive picture, at once reverent, assured and unafraid, of nature and man's relations with it.

These facts and ideas have an important bearing on the so-called race question and the problem of racial equality. I should rather say racial *inequality*, for up till quite recently the naïve belief in the natural inequality of races and people in general, and the inherent superiority of one's own race or people in particular, has almost universally prevailed.

To demonstrate the way in which this point of view permeated even nineteenth-century scientific thought, it is worth recalling that it was largely subscribed to by Darwin in his comments on the Fuegians in the *Voyage of the Beagle* and in more general but more guarded terms in the *Descent of Man*: and Galton himself, against a similar background of travels among backward tribes and on the basis of his own rather curious method of assessment, concluded that different races had achieved different genetic standards, so that, for instance, "the average standard of the Negro race is two grades below our own". This type of belief, after being given a pseudo-scientific backing by non-biological theoreticians like Gobineau and Houston Stewart Chamberlain, was used to justify the Nazis' "Nordic" claims to world domination and their horrible campaign for the extermination of the "inferior, non-Aryan race" of Jews, and is still employed, with support from Holy Writ and the majority of the Dutch Reformed Church in South Africa, to sanction Verwoerd's denial of full human rights to non-whites.

Later investigation has conclusively demonstrated first, that there is no such thing as a "pure race". Secondly, that the obvious differences in level of achievement between different peoples and ethnic groups are primarily cultural, due to differences not in genetic equipment but in historical and environmental opportunity. And thirdly, that when the potentialities of intelligence of two major "races" or ethnic groups, such as whites and negroes or Europeans and Indians, are assessed as scientifically as possible, the frequency curves for the two groups overlap over almost the whole of their extent, so that approximately half the population of either group is genetically stupider (has a lower genetic I.Q.) than the genetically more intelligent half of the other. There are thus large differences in genetic mental endowment *within* single racial groups, but minimal ones *between* different racial groups.

Partly as a result of such studies, but also of the prevalent environmentalist views of Marxist and Western liberal thought, an anti-genetic view has recently developed about race. It is claimed that though ethnic groups obviously differ in physical characters, and that some of these, like pigmentation, nasal form, and number of sweat-glands, were originally adaptive, they do not (and sometimes even that they cannot) differ in psychological or mental characters such as intelligence, imagination, or even temperament.[1]

Against this new pseudo-scientific racial naïveté, we must set the following scientific facts and principles. First, it is clear that the major human races originated as geographical subspecies of *Homo sapiens*, in adaptation to the very different environments to which they had become restricted. Later, of course, expansion and migration reversed the process of differentiation and led to an increasing degree of convergence by crossing, though considerable genetic differentiation remains. Secondly, as Professor Muller has pointed out, it is theoretically inconceivable that such marked physical differences as still persist between the main racial groups should not be accompanied by genetic differences in temperament and mental capacities, possibly of considerable

[1] There is the further point that races may differ considerably in body-build and that Sheldon and others have made it highly probable that body-build is correlated with temperament. Unfortunately, racial differences in body-build have not yet been analysed in terms of Sheldon's somatotypes: here is an important field for research.

extent. Finally, as previously explained, advance in cultural evolution is largely and increasingly dependent on exceptionally well-endowed individuals. Thus two racial groups might overlap over almost the whole range of genetic capacity, and yet one might be capable of considerably higher achievement, not merely because of better environmental and historical opportunity, but because it contained say 2 instead of 1 per cent. of exceptionally gifted men and women. So far as I know, proper scientific research on this subject has never been carried out, and possibly our present methods of investigation are not adequate for doing so, but the principle is theoretically clear and is of vital practical importance.[1]

This does not imply any belief in crude racism, with its unscientific ascription of natural and permanent superiority or inferiority to entire races. As I have just pointed out, approximately half of any large ethnic group, however superior its self-image may be, is in point of fact genetically inferior to half of the rival ethnic group with which it happens to be in social or economic competition and which it too often stigmatizes as permanently and inherently lower. Furthermore, practical experience demonstrates that every so-called race, however underdeveloped its economic and social system may happen to be, contains a vast reservoir of untapped talent, only waiting to be elicited by a combination of challenging opportunity, sound educational methods, and efficient special training. I recently attended an admirable symposium on nutrition in Nigeria where the scientific quality of the African contributions was every whit as high as that of the whites; and African politicians can be just as statesmanlike (and also just as unscrupulously efficient in the political game) as their European or American counterparts.

[1] On the supposition that genetic intelligence is multifactorially (polygenically) determined and that its distribution follows a normal symmetrical curve, it can be calculated that the raising of the mean genetic I.Q. of a population by $1\frac{1}{2}$ per cent. would result in a 50 per cent. increase in the number of individuals with an I.Q. of 160 or over. The proportion of such highly-endowed individuals would rise from about 1 in 30,000 of the total population to about 1 in 20,000. Sir Cyril Burt informs me that if, as is possible, some types of high genetic intelligence are determined by single genes, the increase might be still greater.

The basic fact about the races of mankind is their almost total overlap in genetic potentialities. But the most significant fact for eugenic advance is the large difference in achievement made possible by a small increase in the number of exceptional individuals.

The human species is in desperate need of genetic improvement if the whole process of psychosocial evolution which it has set in train is not to get bogged down in unplanned disorder, negated by over-multiplication, clogged up by mere complexity, or even blown to pieces by obsessional stupidity. Luckily it not only *must* but *can* be improved. It can be improved by the same type of method that has secured the improvement of life in general—from protozoan to protovertebrate, from protovertebrate to primate, from primate to human being—the method of multi-purpose selection directed towards greater achievement in the prevailing conditions of life.

On the other hand, it can *not* be improved by applying the methods of the professional stock-breeder. Indeed the whole discussion of eugenics has been bedevilled by the false analogy between artificial and natural selection. Artificial selection is intensive special-purpose selection, aimed at producing a particular excellence, whether in milk-yield in cattle, speed in race-horses, or a fancy image in dogs. It produces a number of specialized pure breeds, each with a markedly lower variance than the parent species. Darwin rightly drew heavily on its results in order to demonstrate the efficacy of selection in general. But since he never occupied himself seriously with eugenics he did not point out the irrelevance of stock-breeding methods to human improvement. In fact, they are not only irrelevant, but would be disastrous. Man owes much of his evolutionary success to his unique variability. Any attempt to improve the human species must aim at retaining this fruitful diversity, while at the same time raising the level of excellence in all its desirable components, and remembering that the selectively evolved characters of organisms are always the results of compromise between different types of advantage, or between advantage and disadvantage.

Natural selection is something quite different. To start with, it is a shorthand metaphorical term coined by Darwin to denote the teleonomic or directive agencies affecting the process of evolution in nature, and operating through the differential survival and reproduction of genetical variants. It may operate between conspecific individuals, between conspecific populations, between related species, between higher taxa such as genera and families, or between still larger groups of different organizational type, such as Orders and Classes. It may also operate between predator and prey, between parasite and host, and between different synergic assemblages of species, such as symbiotic partnerships and ecological communities. It is in fact universal in its occurrence, though multiform in its mode of action.

Some over-enthusiastic geneticists appear to think that natural selection acts directly on the organism's genetic outfit or genotype. This is not so. Natural selection exerts its effects on animals and plants as working mechanisms: it can operate only on the actualized individual organizations of living substance that we call phenotypes. Its evolutionary action in transforming the genetic outfit of a species is indirect, and depends on the simple fact pointed out by Darwin in the *Origin* that much variation is heritable—in modern terms, that there is a high degree of correlation between phenotypic and genotypic variance. The correlation, however, is never complete, and there are many cases where it is impossible without experimental analysis to determine whether a variant is modificational, due to alteration in the environment, or mutational, due to alteration in the genetic outfit. In certain cases, environmental treatment will produce so-called phenocopies which are indistinguishable from mutants in their visible appearance.

This last fact has led Waddington to an important discovery—the fact that an apparently Lamarckian mode of evolutionary transformation can be precisely simulated by what he calls genetic assimilation. To take an actual example, the rearing of fruitfly larvae on highly saline media produces a hypertrophy of their salt-excreting glands through direct modification. But if selection is practised by breeding from those individuals which show the maximum hypertrophy of their glands, then after some ten or twelve generations, individuals with somewhat hypertrophied glands appear even in cultures on non-saline media. The species has a genetic predisposition, doubtless brought by selection in the past, to react to saline conditions by glandular hypertrophy. The action of the major genes concerned in reactions of this sort can be enhanced (or inhibited) by so-called modifying genes of minor effect. Selection has simply amassed in the genetic outfit an array of such minor enhancing genes strong enough to produce glandular hypertrophy even in the absence of any environmental stimulus. It is only pseudo-Lamarckism, but no less important for that—a significant addition to the theoretical armoury of evolutionary science.

I repeat that the most important effect achieved by natural selection is biological improvement. As G. G. Simpson reminds us, it does so opportunistically, making use of whatever new combination of existing mutant genes, or less frequently of whatever new mutations, happens to confer differential survival value on its possessors. We know of numerous cases where phenotypically identical and adaptive transformations have been produced by different genes or gene-combinations.

Here I must digress a moment to discuss the concept of evolutionary fitness. The biological *avant garde* has chosen to define *fitness* as "net reproductive advantage", to use the actual words employed by Professor Medawar in his Reith Lectures on *The Future of Man*. Any strain of animal, plant, or man which leaves slightly more descendants capable of reproducing themselves than another, is then defined as "fitter". This I believe to be an unscientific and misleading definition. It disregards all scientific conventions as to priority, for it bears no resemblance to what Spencer implied or intended by his famous phrase the *survival of the fittest*.[1] It is also nonsensical in every context save the limited field of population-genetics. In biology, fitness must be defined, as Darwin did with improvement, "in relation to the con-

[1] Darwin did not use the phrase in the first edition of the *Origin of Species*, though in later editions he added it as an equivalent to natural selection.

ditions of life"—in other words, in the context of the general evolutionary situation. I shall call it *evolutionary fitness*, in contradistinction to the purely reproductive fitness of the evangelists of geneticism, which I prefer to designate by the descriptive label of *net* or *differential reproductive advantage*. I expect that the conflict will be resolved by a clarification of selection theory, with acceptance of the fact that there are two very different forms of selection in nature—survival selection operating by the differential survival of phenotypes, and reproductive selection operating by the differential survival of offspring. This will need a revision of our terminology, including a changed definition of fitness.

Meanwhile, I have a strong suspicion that the genetical *avant garde* of today will become the rearguard of tomorrow. In my own active career I have seen a reversal of this sort in relation to natural selection and adaptation. During the first two decades of this century the biological *avant garde* dismissed topics such as cryptic or mimetic coloration, and indeed most discussion of adaptation, as mere "armchair speculation", and played down the role of natural selection in evolution, as against that of large and random mutation. Bateson's enthusiasm rebounded from his early protest against speculative phylogeny into the far more speculative suggestion made *ex cathedra* at a British Association meeting, that all evolution, whether of higher from lower, or of diversity from uniformity, had been brought about by loss mutations; and the great American geneticist, T. H. Morgan, once permitted himself to state in print that, if natural selection had never operated, we should possess all the organisms that now exist and a great number of other types as well! This anti-selectionist *avant garde* of fifty years back has now come over *en masse* into the selectionist camp, leaving only a few retreating stragglers to deliver some rather ineffective parthian shots at their opponents.

Natural selection is a teleonomic or directional agency. It utilizes the inherent genetic variability of organisms provided by the raw material of random mutation and chance recombination, and it operates by the simple mechanism of differential reproductive advantage. But on the evolutionary time-scale it produces biological improvement, resulting in a higher total and especially a higher upper level of evolutionary fitness, involving greater functional efficiency, higher degrees of organization, more effective adaptation, better self-regulating capacity, and finally more mind—in other words an enrichment of qualitative awareness coupled with more flexible behaviour.

Man almost certainly has the largest reservoir of genetical variance of any natural species: selection for the differential reproduction of desirable permutations and combinations of the elements of this huge variance could undoubtedly bring about radical improvement in the human organism, just as it has in prehuman types. But the agency of human transformation cannot be the blind and automatic natural selection of the prehuman sector. That, as I have already stressed, has been relegated to a subsidiary role in the human phase of evolution. Some form of psychosocial selection is needed, a selection as non-natural as are most human activities, such as wearing clothes, going to war, cooking food, or employing arbitrary systems of communication. To be effective, such "non-natural" selection must be conscious, purposeful and planned. And since the tempo of cultural evolution is many thousands of times faster than that of biological transformation, it must operate at a far higher speed than natural selection if it is to prevent disaster, let alone produce improvement.

Luckily there is today at least the possibility of meeting both these prerequisites: we now possess an accumulation of established knowledge and an array of tested methods which could make intelligent, scientific and purposeful planning possible. And we are in the process of discovering new techniques which could raise the effective speed of the selective process to a new order of magnitude. The relevant new knowledge mainly concerns the various aspects of the evolutionary process—the fact that there are no absolutes or all-or-nothing effects in evolution and that all organisms and all their phenotypic characters represent a compromise or balance between competing advantages and disadvantages; the effect of selection on populations in different environmental conditions; the origin of adaptation; and the general improvement of different evolutionary lines in relation to the conditions of their life. The notable new techniques include effective methods of birth-control, the successful development of grafted fertilized ova in new host-mothers, artificial insemination, and the conservation of function in deep-frozen gametes.

We must first keep in mind the elementary but often neglected fact that the characters of organisms which make for evolutionary success or failure, are not inherited as such. On the contrary, they develop anew in each individual, and are always the resultant of an interaction between genetic determination and environmental modification. Biologists are often asked whether heredity or environment is the more important. It cannot be too often emphasized that the question should never be asked. It is as logically improper to ask a biologist to answer it as it is for a prosecuting counsel to ask a defendant when he stopped beating his wife. It is the phenotype which is biologically significant and the phenotype is a resultant produced by the complex interaction of hereditary and environmental factors. Eugenics, in common with evolutionary biology in general, needs this phenotypic approach.

Man's evolution occurs on two different levels and by two distinct methods, the genetic, based on the transmission and variation of genes and gene-combinations, and the psycho-social or cultural, based on the transmission and variation of knowledge and ideas.

Professor Medawar, in his Reith Lectures on *The Future of Man*, while admitting in his final chapter that man possesses "a new, non-genetical, system of heredity and evolution" claims that this is "a new kind of biological evolution (I emphasize, a biological evolution)". I must insist that this is incorrect. The psychosocial process—in other words, evolving man—is a new *state* of evolution, a new *phase* of the cosmic process, as radically different from the prehuman biological phase as that is from the inorganic or prebiological phase; and this fact has important implications for eugenics.

An equally elementary but again often neglected fact is

that organisms are not significant—in plain words, are meaningless—except in relation to their environment. A fish is not a thing-in-itself: it is a type of organism evolved in relation to an active life in large or medium-sized bodies of water. A cactus has biological significance only in relation to an arid habitat, a woodpecker only in relation to an arboreal one. Man, however, is in a unique situation. He must live not only in relation with the physico-chemical and biological environment provided by nature, but with the psychosocial environment of material and mental habitats which he has himself created.

Man's psychosocial environment includes his beliefs and purposes, his ideals and his aims: these are concerned with what we may call the habitat of the future, and help to determine the *direction* of his further evolution. All evolution is directional and therefore relative. But whereas the direction of biological evolution is related to the continuing improvement of organisms in relation to their conditions of life, human evolution is related to the improvement of the entire psychosocial process, including the human organism, in relation to man's purposes and beliefs, long-term as well as short-term. Only in so far as those purposes and beliefs are grounded on scientific and tested knowledge, will they serve to steer human evolution in a desirable direction. In brief, biological evolution is given direction by the blind and automatic agency of natural selection operating through material mechanisms, human evolution by the agency of psychosocial guidance operating with the aid of mental awareness, notably the mechanisms of reason and imagination.

To be effective, such awareness must clearly be concerned with man's environmental situation as well as his genetic equipment. Twenty-five years ago, in my first Galton Lecture, I pointed out the desirability of eugenists relating their policies to the social environment. Today I would go further, and stress the need for planning the environment in such a way as will promote our eugenic aims. By 1936, it was already clear that the net effect of present-day social policies could not be eugenic, and was in all probability dysgenic.

But, as Muller has demonstrated, this was not always so. In that long period of human history during which our evolving and expanding hominid ancestors lived in small and tightly knit groups competing for territorial and technological success, the social organization promoted selection for intelligent exploration of possibilities, devotion and co-operative altruism: the cultural and the genetic systems reinforced each other. It was only much later, with the growth of bigger social units of highly organized civilizations based on status and class differentials, that the two became antagonistic; the sign of genetic transformation changed from positive to negative and definite genetic improvement and advance began to halt, giving way to the possibility and later the probability of genetic regression and degeneration.

This probability has been very much heightened during the last century, partly by the differential multiplication of economically less favoured classes and groups in many parts of the world, partly by the progress of medicine and public health, which has permitted numbers of genetically defective human beings to survive and reproduce; and today it has been converted into a certainty by the series of atomic and nuclear explosions which have been set off since the end of the last war. There is still dispute as to the degree of damage this has done to man's genetic equipment. There can be no dispute as to the fact of damage: any addition to man's load of mutations can only be deleterious, even if a few of them may possibly come to be utilized in neutral or even favourable new gene-combinations.

Now that we have realized these portentous facts, it is up to us to reverse the process and to plan a society which will favour the increase instead of the decrease of man's desirable genetic capacities for intelligence and imagination, empathy and co-operation, and a sense of discipline and duty.

The first step must be to frame and put into operation a policy designed to reduce the rate of human increase before the quantitative claims of mere numbers override those of quality and prevent any real improvement, social and economic as much as eugenic. I would prophesy that within a quite short time, historically speaking, we shall find ourselves aiming at an absolute reduction of the population in the world in general, and in overcrowded countries like Britain, India and China, Japan, Java and Jamaica in particular; the quantitative control of population is a necessary prerequisite for qualitative improvement, whether psychosocial or genetic.

Science seems to be nearing a breakthrough to cheap and simple methods of birth-control, or reproduction-control as it should more properly be called. The immediate needs are for much-increased finance for research, testing, pilot projects, motivation studies and the education of public opinion, and an organized campaign against the irrational attitudes and illiberal policies of various religious and political organizations. Simultaneously responsible opinion must begin to think out ways in which social and economic measures can be made to promote desirable genetic trends and reproductive habits.

Many countries have instituted family allowance systems which are not graded according to number of children, and some, like France, even provide financial inducements which encourage undesirably large families. It should be easy to devise graded family allowance systems in which the allowances for the first two or three children would be really generous, but those for further children would rapidly taper off. In India, there have even been proposals to tax parents for children above a certain number, and in some provinces, men fulfilling certain conditions are paid to be vasectomized.

A powerful weapon for adequate population-control is ready to the hand of the great grant-giving and aid-providing agencies of the modern world—international agencies such as the U.N. and its Technical Assistance Board representing its various Specialized Agencies like F.A.O. and Unesco, the World Bank and the International Finance Corporation Administration; national agencies like the Colombo Plan and the Inter-American Development Fund; and the great private Foundations (wittily categorized as *philanthropoid* by that remarkable man Frederick Keppel) like Rockefeller and Ford, Gulbenkian, Nuffield and Carnegie.

At the moment, much of the financial and technical aid provided by these admirable bodies is being wasted by being flushed down the drain of excess population instead of into the channels of positive economic and cultural development, or is even defeating its own ends by promoting excessive and over-rapid population-increase. Bankers do not make loans unless they are satisfied of the borrower's credit-worthiness. Surely these powerful agencies, public or private, should not provide loans or grants or other aid unless they are satisfied of the recipient nation's demographic credit-worthiness.

At last I reach my specific subject—eugenics, with its two aspects, negative and positive. Negative eugenics aims at preventing the spread and especially the increase of defective or undesirable human genes or gene-combinations, positive eugenics at securing the reproduction and especially the increase of favourable or desirable ones.[1]

Negative eugenics has become increasingly urgent with the increase of mutations due to atomic fallout, and with the increased survival of genetically defective human beings, brought about by advances in medicine, public health, and social welfare. But it must, of course, attempt to reduce the incidence, or the manifestation, of every kind of genetic defect. Such defects include high genetic proneness to diseases such as diabetes, schizophrenia (which affects 1 per cent. of the entire human population), other insanities, myopia, mental defect and very low I.Q., as well as more clear-cut defects like colour-blindness or haemophilia.

When defects depend on a single dominant gene, as with Huntington's chorea, transmission can of course be readily prevented by persuading the patient to refrain from reproducing himself. With sexlinked defects like haemophilia, Duchenne-type muscular dystrophy, or HCN "smell-blindness", this will help, but the method should be supplemented by counselling his sisters against marriage. This will be more effective and more acceptable when, as seems possible, we can distinguish carriers heterozygous for

---

[1] In the past, these aims have been generally expressed in terms of defective or desirable *stocks* or *strains*. With the progress of genetics, it is better to reformulate them in terms of genes and gene-combinations.

the defect from non-carriers. This is already practicable with some autosomic recessive defects, notably sickle-cell anaemia. Here, registers of carriers have been established in some regions, and they are being effectively advised against inter-marriage. This will at least prevent the manifestation of the defect. The same could happen with galactosaemia, and might be applicable to relatives of patients with defects like phenylketonuria and agammoglobulinaemia.

In addition, the marked differential increase of lower-income groups, classes and communities during the last hundred years cannot possibly be eugenic in its effects. The extremely high fertility of the so-called social problem group in the slums of industrial cities is certainly anti-eugenic.

As Muller and others have emphasized, unless these trends can be checked or reversed, the human species is threatened with genetic deterioration, and unless this load of defects is reduced, positive eugenics cannot be successfully implemented. For this we must reduce the reproduction rate of genetically defective individuals: that is negative eugenics.

The implementation of negative eugenics can only be successful if family planning and eugenic aims are incorporated into medicine in general and into public health and other social services in particular. Its implementation in practice will depend on the use of methods of contraception or sterilization, combined where possible with A.I.D. (artificial insemination by donor) or other methods of vicarious parenthood. In any case, negative eugenics is of minor evolutionary importance and the need for it will gradually be superseded by efficient measures of positive eugenics.

In cases of specific genetic defect, voluntary sterilization is probably the best answer.[1] In the defective married male, it should be coupled with artificial parenthood (A.P.) by donor insemination (A.I.D.) as the source of children. In the defective female, the fulfilments of child-rearing and family life will have to be secured by adoption until such time—which may not be very distant—as improved technique makes possible artificial parenthood by transfer of fertilized

[1] It will be even more satisfactory if, as now appears likely, reversible male sterilization (vasectomy) becomes practicable.

ova, which we may call A.O.D. In both cases, it must be remembered that sterilization does not prevent normal healthy and happy sexual intercourse.

Certified patients are now prevented from reproducing themselves by being confined in mental hospitals. If sterilized, they might be allowed to marry if this were considered likely to ameliorate their condition.

In the case of the so-called social problem group, somewhat different methods will be needed. By social problem group I mean the people, all too familiar to social workers in large cities, who seem to have ceased to care, and just carry on the business of bare existence in the midst of extreme poverty and squalor. All too frequently they have to be supported out of public funds, and become a burden on the community. Unfortunately they are not deterred by the conditions of existence from carrying on with the business of reproduction: and their mean family size is very high, much higher than the average for the whole country.

Intelligence and other tests have revealed that they have a very low average I.Q.; and the indications are that they are genetically subnormal in many other qualities, such as initiative, pertinacity, general exploratory urge and interest, energy, emotional intensity, and will-power. In the main, their misery and improvidence is not their fault but their misfortune: our social system provides the soil on which they can grow and multiply, but with no prospects save poverty and squalor.

Here again, voluntary sterilization could be useful. But our best hope, I think, must lie in the perfection of new, simple and acceptable methods of birth-control, whether by an oral contraceptive or perhaps preferably by immunological methods involving injections. Compulsory or semi-compulsory vaccination, inoculation and isolation are used in respect of many public health risks: I see no reason why similar measures should not be used in respect of this grave problem, grave both for society and for the unfortunate people whose increase has been actually encouraged by our social system.

Many social scientists and social workers in the West, as

270

well as all orthodox Marxists, are environmentalists. They seem to believe that all or most human defects, including many that Western biologists would regard as genetic, can be dealt with, cured or prevented by improving social environment and social organization. Even some biologists, like Professor Medawar, agree in general with this view, though he admits a limited role for negative eugenics, in the shape of what he calls "genetic engineering". For him, the "newer solution" of the problem, which "goes some way towards making up for the inborn inequalities of man", is simply to improve the environment. With this I cannot agree. Although certain particular problems can be dealt with in this way, for instance proneness to tuberculosis by improving living conditions and preventing infection, such methods cannot cope with the general problem of genetic deterioration, because this, if not checked, will steadily increase through the accumulation of mutant genes which otherwise would have been eliminated.

It is true that many diseases or defects with a genetic basis, like diabetes or myopia, can be cured by treatment, though almost always with some expense, trouble or discomfort to the defective person as well as to society. But if the incidence of such defects (not to mention the many others for which no cure or remedy is now known) were progressively multiplied, the burden would grow heavier and heavier and eventually wreck the social system. As in all other fields, we need to combine environmental and genetic measures, and if possible render them mutually reinforcing.

Against the threat of genetic deterioration through nuclear fallout there are only two courses open. One is to ban all nuclear weapons and stop bomb-testing; the other is to take advantage of the fact that deep-frozen mammalian sperm will survive, with its fertilizing and genetic properties unimpaired, for a long period of time and perhaps indefinitely, and accordingly to build deep shelters for sperm-banks— collections of deep-frozen sperm from a representative sample of healthy and intelligent males. A complete answer must wait for the successful deep-freezing of ova also. But this may be achieved in the fairly near future, and in any case shelters for sperm-banks will give better genetic results than shelters for people, as well as being very much cheaper.

Positive eugenics has a far larger scope and importance than negative. It is not concerned merely to prevent genetic deterioration, but aims to raise human capacity and performance to a new level.

For this, however, it cannot rely on measures designed to produce merely a slight differential increase of genetically superior stocks, generation by generation. This is the way natural selection obtains its results, and it worked all right during the biological phase, when immense spans of time were available. But with the accelerated tempo of modern psychosocial evolution, much quicker results are essential. Luckily modern science is providing the necessary techniques, in the shape of artificial insemination and the deep-freezing of human gametes. The effects of superior germ-plasm can be multiplied ten or a hundredfold through the use of what I call E.I.D.—eugenic insemination by deliberately preferred donors—and many thousandfold if the superior sperm is deep-frozen.

This multiplicative method, harnessing man's deep desires for a better future, was first put forward by H. J. Muller and elaborated by Herbert Brewer, who invented the terms *eutelegenesis* and *agapogeny* for different aspects of it. Some such method, or what we may term Euselection—deliberate encouragement of superior genetic endowment—can produce immediate results. Couples who adopt this method of vicarious parenthood can be rewarded by children outstanding in qualities admired and preferred by the couple themselves.

When deep-frozen ova too can be successfully engrafted into women, the speed and efficiency of the process could of course be intensified.

Various critics insist on the need for far more detailed knowledge of genetics and selection before we can frame a satisfactory eugenic policy or even reach an understanding of evolution. I can only say how grateful I am that neither Galton nor Darwin shared these views, and state my own firm belief that they are not valid. Darwin knew nothing—I repeat *nothing*—about the actual mechanism of biological variation and inheritance: yet he was possessed of what I can only call a common-sense genius which gave him a general understanding of the biological process and enabled him to frame a theory of the process whose core remains unshaken and which has been able successfully to incorporate all the modifications and refinements of recent field study and genetic experiment.

Neither did the automatic process of natural selection "know" anything about the mechanisms of evolution. Luckily this did not prevent it from achieving a staggering degree of evolutionary transformation, including miracles of adaptation and improvement. From his seminal idea, Darwin was able to deduce important general principles, notably that natural selection would automatically tend to produce both diversification (adaptive radiation) and improvement (biological advance or progress) in organization, but that lower types of organization would inevitably survive alongside higher.

Critics of positive eugenics like Medawar inveigh against what they call "*geneticism*". However, he himself is guilty on this count, for he has accepted the population geneticists' claim (which I have discussed earlier) that theirs is the only scientifically valid definition of *fitness*; and this in spite of his admission that one organic type can be more "advanced" than another, and that "human beings are the outcome of a process which can perfectly well be described as an advancement". However, he equates advancement with mere increase in complexity of the "genetical instructions" given to the animal: if he had thought in broad evolutionary instead of restricted genetic terms he would have seen that biological advance involves improved organization of the phenotype; that fitness in this geneticismal sense is a purely reproductive fitness; and that we must also take into account immediate phenotypic fitness and long-term evolutionary fitness. To put it in a slightly different way, "fitness" as measured by differential survival of offspring is merely the mechanism by which the long-term improvement of true biological fitness is realized.

273

Recent genetic studies have shown the wide-spread occurrence of genetic polymorphism, in animal species and man, whether in the form of sharply distinct morphs (as with colour-blindness and other sensory morphisms), in multiplicity of slightly different alleles, or merely in a very high degree of potential variance. Some critics of positive eugenics maintain that this state of affairs will prevent or at least strongly impede any large-scale genetic improvement, owing to the resistance to change offered by genetic polymorphisms maintained by means of heterozygote advantage, which appear to comprise the majority of polymorphic systems.

It has further been suggested, notably by Professor Penrose, that people heterozygous for genes determining general intellectual ability, and therefore of medium or mediocre intelligence, are reproductively "fitter"—more fertile—than those of high or low intelligence, and accordingly that, as regards genetic intelligence, the British population is in a state of natural balance. If so, it would be difficult to try to raise its average level by deliberate selective measures, and equally difficult for the level to sink automatically as the result of differential fertility of the less intelligent groups.

Although Medawar appears to disagree with Penrose's main contention, he concludes that: "If a tyrant were to attempt to raise the intelligence of all of us to its present maximum . . . I feel sure that his efforts would be self-defeating: the population would dwindle in numbers and, in the extreme case, might die out." It is true that he later enters a number of minor caveats, but his main conclusion remains. This to me appears incomprehensible. If selection has operated, as it certainly has done in the past, during the passage from Pithecanthropus to present-day man, to bring about a very large rise in the level of genetic intelligence, why can it not bring about a much smaller rise in the immediate future? There are no grounds for believing that modern man's system of genetic variance differs significantly from that of his early human ancestors.

As regards balanced morphisms, it is of course true that they constitute stable elements in an organism's genotype. However, when their stability is mainly due to linkage with a lethal, and therefore to double-dose disadvantage rather than to heterozygote advantage, they may be destabilized by breaking the linkage. In any case, morphisms stable in one environment may sometimes be broken up in another. This has happened, for instance, with the white-yellow sex-limited morphism of the butterfly *Colias eurythema*, which in high latitudes has ceased to exist, and the local population is monomorphic, all homozygous white.

Certainly some morphisms show very high stability. For instance the PTC (phenylthiocarbamide) taste morphism occurs in apparently identical form both in chimpanzees and man. The majority of both species find PTC very disagreeably bitter, while about 25 per cent. cannot taste it at all. Accordingly this morphic system must have resisted change for something like ten million years. However, this remarkable stability of a specific genotypic component of the primate stock has not prevented the transformation of one branch of that stock into man!

Similar arguments apply to linked polygenic systems and to the general heterozygosity in respect of small allelic differences shown by so many organisms, including man. In the former case, Mather has shown how selection can break the linkage and make the frozen variability available for new recombinations and new evolutionary change. In the latter case, the stability need not be so intense as with clear-cut morphisms.

Frequently, it appears, polymorphism depends not so much on heterotic advantage as on a varying balance of advantage between the alleles concerned in different conditions: one allele is more advantageous in certain conditions, another in other conditions. The polymorphism is therefore a form of insurance against extreme external changes and gives flexibility in a cyclically or irregularly varying environment. Loose polymorphic systems of this general type can readily be modified by the incorporation of new and the elimination of old mutant alleles and the incorporation of new ones in response to directional changes in environment. In any case, their widespread existence has not stood in the way of directional evolutionary change, including the transformation of a protohominid into man. Why should they stand in the way of man's further genetic evolution?

The same reasoning applies to those numerous cases where high genetic variance, actual or potential, is brought about by multiple genic polymorphism, when many genes of similar action exist, often in a number of slightly different allelic forms.

In all these cases the critics of eugenics have been guilty of that very "geneticism" which they deplore. They approach the subject from the standpoint of population-genetics. If they were to look at it from an evolutionary standpoint, their difficulties would evaporate, and they would see that their objections could not be maintained.

Two further objections are often made to positive eugenics. One is by way of a question—who is to decide which type to select for? The other, which is by way of an answer to the first, is to assert that effective selection needs authoritarian methods and can only be put into operation by some form of dogmatic tyranny, usually stigmatized as intolerable or odious.

Both these objections reveal the same lack of understanding of psychosocial evolution as the genetical objections revealed about biological evolution: more simply, they demonstrate the same lack of faith in the potentialities of man that the purely genetical objections showed in the actual operative realizations of life.

For one thing, dogmatic tyranny in the modern world is becoming increasingly self-defeating: partly because it is dogmatic and therefore essentially unscientific, partly because it is tyrannical and therefore in the long run intolerable. But the chief point is that human improvement never works solely or even mainly by such methods and is doing so less and less as man commits himself more thoroughly to the process of general self-education.

Let me take an example. Birth-control resembles eugenics in being concerned with that most violent arouser of emotion and prejudice, human reproduction. However, during my own career, I have witnessed the subject break out of the dark prison of taboo into the international limelight. It was only in 1917 that Margaret Sanger was given a jail sentence for disseminating birth-control information. In the late 'twenties, when I was already over 40, I was summoned before the first Director-General of the B.B.C., now Lord Reith, and rebuked for having contaminated the British ether with such a shocking subject. Yet two years ago an international gathering in New York paid tribute to Margaret Sanger as one of the great women of our time: *Time* and *Life* magazines both published long and reasoned articles on how to deal with the population explosion, and two official U.S. committees reported in favour of the U.S. conducting more research on birth-control methods and even of giving advice on the subject if requested by other nations. And today one can hardly open a copy of the most respectable newspapers without finding at least one reference to the grievous effects of population-increase and population-density on one or another aspect of human life in one or another country of the globe, including our own. Meanwhile, six nations have started official policies of family planning and population control, and many others are unofficially encouraging them.

Birth-control, in fact, has broken through—and in so doing it has changed its character and its methods. It began as a humanitarian campaign for the relief of suffering human womanhood, conducted by a handful of heroic figures, mostly women. It has now become an important social, economic and political campaign, led by powerful private associations, and sometimes the official or semi-official concern of national governments. Truth, in fact, prevails—though its prevailing demands time, public opprobrium of the self-sacrificing pioneers at the outset, and public discussion, backed by massive dissemination of facts and ideas, to follow.

We can safely envisage the same sort of sequence for evolutionary eugenics, operating by what I have called euselection, though doubtless with much difference in detail. Thus the time to achieve public breakthrough might be longer because the idea of Euselection by delegated paternity runs counter to a deep-rooted sense of proprietary parenthood. On the other hand it might be shorter, since there is such a rapid increase in the popular understanding of science and in the agencies of mass communication and information, and above all because of the profound dissatisfaction with traditional ideas and social systems, which portends the drastic recasting of thought and attitude that I call the Humanist Revolution.

Some things, at least, are clear. First, we need to establish the legality, the respectability, and indeed the morality of A.I.D. It must be cleared of the stigma of sin ascribed to it by Church dignitaries like Lord Fisher when Archbishop of Canterbury, and from the legal difficulties to its practice raised by the lawyers and administrators. Most importantly, the notion of donor secrecy must be abolished. Parents desiring A.I.D. should have not only the right but the duty of choice. For the time being, it may possible be best that the name and personal identity of a donor should not be known to the acceptors, but there should certainly be a register of certified donors kept by medical men (and I would hope by the National Health Service) which would give particulars of their family histories. This would enable acceptors to exert a degree of conscious selection in choosing the father of the child they desire, and so pave the way for the supersession of blind and secrecy-ridden A.I.D. by an open-eyed and proudly accepted E.I.D. where the E stands for *Eugenic*.

The pioneers of E.I.D., whether its publicists or its practitioners, will undoubtedly suffer all kinds of abusive prejudice—they will be accused of mortal sin, of theological impropriety, of immoral and unnatural practices. But they can take heart from what has happened in the field of birth-control, and can be confident that the rational control of reproduction aimed at the prevention of human suffering and frustration and the promotion of human well-being and fulfilment will in the not too distant future come to be recognized as a moral imperative.

The answers to the questions I mentioned at the beginning of this section are now, I hope, clear. There will be no single type to be selected for, but a range of preferred types; and this will not be chosen by any single individual or committee. The choice will be a collective choice representing the varied preferences and ideals of all the couples practising euselection by E.I.D., and it will not be dogmatically imposed by any authoritarian agency, though as general acceptance of the method grows, it will be reinforced by public opinion and official leadership. The way is open for the most significant step in the progress of mankind—the deliberate improvement of the species by scientific and democratic methods.

All the objections of principle to a policy of positive eugenics fall to the ground when the subject is looked at in the embracing perspective of evolution, instead of in the limited perspective of population-genetics or the short-term perspective of existing socio-political organization. Meanwhile the obvious practical difficulties in the way of its execution are being surmounted, or at least rendered surmountable, by scientific discovery and technical advance.

In evolutionary perspective, eugenics—the progressive genetic improvement of the human species—inevitably takes its place among the major aims of evolving man. What should we eugenists do in the short term to promote this long-term aim? We must of course continue to do and to encourage research on human genetics and reproduction, including methods of conception-control and sterilization. The establishment of the Darwin Research Fellowships by the Eugenic Society is an important milestone in this field: I hope such research activities will rapidly increase.

We must continue to support negative eugenic measures, especially perhaps in respect of the so-called social problem group. We should assuredly continue to be concerned about population-increase, and to support all agencies and organizations aiming at sane and scientific policies of population-control. We must equally support all agencies giving eugenic advice and marriage guidance. Since significant eugenic improvement depends on donor insemination, we must do all we can to win public support for A.I.D., and to improve current practices in the subject.

In general, we must bring home to the general public the possibility of real genetic improvement, the burden it could lift off human shoulders, the hope it could kindle in human hearts. We must make people understand that social and cultural amelioration are not enough. If they are not to turn into temporary palliatives or degenerate into mere environmental tinkering, they must be combined with genetic amelioration, or at least with the hope of it in the future.

To ensure this, not only must the eugenics movement help to educate the public and especially the members of the professions—medical, educational, scientific, administrative, and others—in respect of eugenics, but it must make every effort to get the educational system improved at all levels, so as to provide everyone with the necessary minimum of biological understanding—an understanding of reproduction and population, genetics and selection, ecology and conservation, and above all of the process of evolution in its awe-inspiring sweep and of man's specific significance and responsibility in that comprehensive process.

If, as I firmly believe, man's role is to do the best he can to manage the evolutionary process on this planet and to guide its future course in a desirable direction, fuller realization of genetic possibilities becomes a major motivation for man's efforts, and eugenics is revealed as one of the basic human sciences.

INDEX OF PERSONS

SUBJECT INDEX

284

ORDER FORM
GREAT BOOKS IN PHILOSOPHY PAPERBACK SERIES

ETHICS

Aristotle—*The Nicomachean Ethics*	$8.95
Marcus Aurelius—*Meditations*	5.95
Jeremy Bentham—*The Principles of Morals and Legislation*	8.95
Epictetus—*Enchiridion*	3.95
Immanuel Kant—*Fundamental Principles of the Metaphysic of Morals*	4.95
John Stuart Mill—*Utilitarianism*	4.95
George Edward Moore—*Principia Ethica*	8.95
Friedrich Nietzsche—*Beyond Good and Evil*	8.95
Bertrand Russell On Ethics, Sex, and Marriage (edited by Al Seckel)	17.95
Benedict de Spinoza—*Ethics* and *The Improvement of the Understanding*	9.95

SOCIAL AND POLITICAL PHILOSOPHY

Aristotle—*The Politics*	7.95
The Basic Bakunin: Writings, 1869–1871 (translated and edited by Robert M. Cutler)	10.95
Edmund Burke—*Reflections on the Revolution in France*	7.95
John Dewey—*Freedom and Culture*	10.95
G. W. F. Hegel—*The Philosophy of History*	9.95
Thomas Hobbes—*The Leviathan*	7.95
Sidney Hook—*Paradoxes of Freedom*	9.95
Sidney Hook—*Reason, Social Myths, and Democracy*	11.95
John Locke—*Second Treatise on Civil Government*	4.95
Niccolo Machiavelli—*The Prince*	4.95
Karl Marx/Frederick Engels—*The Economic and Philosophic Manuscripts of 1844* and *The Communist Manifesto*	6.95
John Stuart Mill—*Considerations on Representative Government*	6.95
John Stuart Mill—*On Liberty*	4.95
John Stuart Mill—*On Socialism*	7.95
John Stuart Mill—*The Subjection of Women*	4.95

Thomas Paine—*Rights of Man*   7.95
Plato—*The Republic*   9.95
Plato on Homosexuality: Lysis, Phaedrus, and *Symposium*   6.95
Jean-Jacques Rousseau—*The Social Contract*   5.95
Mary Wollstonecraft—*A Vindication of the Rights of Women*   6.95

METAPHYSICS/EPISTEMOLOGY

Aristotle—*De Anima*   6.95
Aristotle—*The Metaphysics*   9.95
George Berkeley—*Three Dialogues Between Hylas and
   Philonous*   4.95
René Descartes—*Discourse on Method* and *The Meditations*   6.95
John Dewey—*How We Think*   10.95
Sidney Hook—*The Quest for Being*   11.95
David Hume—*An Enquiry Concerning Human Understanding*   4.95
David Hume—*Treatise of Human Nature*   9.95
William James—*Pragmatism*   7.95
Immanuel Kant—*Critique of Pure Reason*   9.95
Gottfried Wilhelm Leibniz—*Discourse on Method* and the
   Monadology   6.95
Plato—*The Euthyphro, Apology, Crito,* and *Phaedo*   5.95
Bertrand Russell—*The Problems of Philosophy*   8.95
Sextus Empiricus—*Outlines of Pyrrhonism*   8.95

PHILOSOPHY OF RELIGION

Ludwig Feuerbach—*The Essence of Christianity*   8.95
David Hume—*Dialogues Concerning Natural Religion*   5.95
John Locke—*A Letter Concerning Toleration*   4.95
Thomas Paine—*The Age of Reason*   13.95
Bertrand Russell On God and Religion (edited by Al Seckel)   17.95

ESTHETICS

Aristotle—*The Poetics*   5.95

GREAT MINDS PAPERBACK SERIES

ECONOMICS

RELIGION

SCIENCE

HISTORY

SPECIAL—For your library . . . the entire collection of 50 "Great Books in Philosophy" and 9 "Great Minds" available at a savings of more than 15%. Only $340.00 for the "Great Books" and $84.00 for the "Great Minds" (plus $12.00 postage and handling). Please indicate "Great Books/Great Minds—Complete Set" on your order form.

The books listed can be obtained from your book dealer or directly from Prometheus Books. Please indicate the appropriate titles. Remittance must accompany all orders from individuals. Please include $3.50 postage and handling for the first book and $1.75 for each additional title (maximum $12.00, NYS residents please add applicable sales tax). Books will be shipped fourth-class book post. **Prices subject to change without notice.**

Send to _____
(Please type or print clearly)

Address _____

City _____ State _____ Zip _____

Amount enclosed _____

Charge my ☐ **VISA** ☐ **MasterCard**

Account # [ ][ ][ ][ ][ ][ ][ ][ ][ ][ ][ ][ ][ ][ ][ ][ ][ ]

Exp. Date _____/_____ Tel.# _____

Signature _____

Prometheus Books Editorial Offices
700 E. Amherst St., Buffalo, New York 14215

Distribution Facilities
59 John Glenn Drive, Amherst, New York 14228

Phone Orders call toll free: (800) 421-0351
FAX: (716) 691-0137
Please allow 3-6 weeks for delivery